Social Epistemology and Epistemic Agency

Collective Studies in Knowledge and Society

Series Editor:
James H. Collier is Associate Professor of Science and Technology in Society at Virginia Tech.

This is an interdisciplinary series published in collaboration with the Social Epistemology Review and Reply Collective. It addresses questions arising from understanding knowledge as constituted by, and constitutive of, existing, dynamic and governable social relations.

Titles in the Series

The Future of Social Epistemology: A Collective Vision edited by James H. Collier
Social Epistemology and Technology: Toward Public Self-Awareness Regarding Technological Mediation edited by Frank Scalambrino
Social Epistemology and Epistemic Agency edited by Patrick J. Reider
Socrates Tenured: The Institutions of 21st Century Philosophy, Adam Briggle and Robert Frodeman (forthcoming)

Social Epistemology and Epistemic Agency

Decentralizing Epistemic Agency

Edited by Patrick J. Reider

ROWMAN &
LITTLEFIELD
———— INTERNATIONAL
London • New York

Published by Rowman & Littlefield International, Ltd.
Unit A, Whitacre Mews, 26-34 Stannary Street, London SE11 4AB
www.rowmaninternational.com

Rowman & Littlefield International, Ltd. is an affiliate of Rowman & Littlefield
4501 Forbes Boulevard, Suite 200, Lanham, Maryland 20706, USA
With additional offices in Boulder, New York, Toronto (Canada), and London (UK)
www.rowman.com

Copyright © 2016 by Patrick J. Reider and Contributors

All rights reserved. No part of this book may be reproduced in any form or by any electronic or mechanical means, including information storage and retrieval systems, without written permission from the publisher, except by a reviewer who may quote passages in a review.

British Library Cataloguing in Publication Information Available
A catalogue record for this book is available from the British Library

ISBN: HB 978-1-78348-347-1
ISBN: PB 978-1-78348-348-8

Library of Congress Cataloging-in-Publication Data

Names: Reider, Patrick J.
Title: : Social epistemology and epistemic agency : decentralizing epistemic agency / edited by Patrick J. Reider.
Description: : Lanham : Rowman & Littlefield International, 2016. | Series: Collective studies in knowledge and society | Includes bibliographical references and index.
Identifiers: : LCCN 2016023766 (print) | LCCN 2016024124 (ebook) | ISBN 9781783483471 (cloth) | ISBN 9781783483488 (pbk.) | ISBN 9781783483495 (electronic)
Subjects: : LCSH: Social epistemology. | Agent (Philosophy) | Epistemics.
Classification: LCC BD175 .S62155 2016 (print) | LCC BD175 (ebook) | DDC 121--dc23 LC record available at https://lccn.loc.gov/2016023766

∞™ The paper used in this publication meets the minimum requirements of American National Standard for Information Sciences Permanence of Paper for Printed Library Materials, ANSI/NISO Z39.48-1992.

Printed in the United States of America

Contents

Introduction: What Is Social Epistemology and Epistemic Agency?   vii
   Patrick J. Reider

I: Anchor Articles

1. A Proposed Research Program for Social Epistemology   3
   Sanford C. Goldberg
2. A Sense of Epistemic Agency Fit for Social Epistemology   21
   Steve Fuller

II: Responses and Further Considerations

3. Two Kinds of Social Epistemology and the Foundations of Epistemic Agency   43
   Finn Collin
4. Fuller's Social Epistemology and Epistemic Agency   61
   Francis Remedios and Valentine Dusek
5. Agency and Disagreement   75
   Paul Faulkner
6. Disciplines, the Division of Epistemic Labor, and Agency   91
   Fred D'Agostino
7. The Distribution of Epistemic Agency   109
   Orestis Palermos and Duncan Pritchard
8. Toward Fluid Epistemic Agency: Differentiating the Terms Being, Subject, Agent, Person, and Self   127
   Frank Scalambrino
9. "Epistemic Agency": A Hegelian Perspective   145
   Angelica Nuzzo
10. Epistemic Agency as a Social Achievement: Rorty, Putnam, and Neo-German Idealism   161
   Patrick J. Reider

Index   179
Author Biographies   181

Introduction

What Is Social Epistemology and Epistemic Agency?

Patrick J. Reider

The aim of this book is to provide a compendium of views that may serve as an introduction and resource for the study of epistemic agency, as framed within the discipline of social epistemology.[1] In introducing the main themes of the book I will identify the divergent positions held towards social epistemology and sketch how these may bring different perspectives to bear on the topic of epistemic agency.

1. KNOWLEDGE AS A CORE HUMAN PROJECT

Many of our interests are peculiar to our own concerns and circumstances, making it difficult to convincingly argue that others *should* hold these interests as well. There are, however, subjects that should concern *all* human beings, in that failure to show interest in them suggests some deficiency in the quality of one's life. For example, it may be the case there are many forms of arts that may not attract one's personal interest, but if one had no interest in any form of art, then one could argue that this might constitute a limitation on one's human capacities. Something similar might be said with respect to *knowledge*.

Not only does knowledge permit culture, history, and a wide range of human practices, it is also necessary for the formation of moral judgments.[2] For that reason, the achievement of knowledge, and all the practices dependent on knowledge, is arguably at the center of what makes us human. Thus the study of knowledge (i.e., epistemology) touches upon a central core of our shared humanity, and ignorance of its possible formulations and means of acquisition, dissemination, and retention limits one's capacity to understand human beings as the moral, cultural, political, and curious beings that we are. If one has interest in humans and what they do, *knowledge* should be at the center of her study of them. As a philosophical endeavor, discerning the manner in which knowledge contributes to our humanity, and more specifically, the manner in which it 1) *ought* to impact human relations and organizations and 2) how it *ought* to be understood, acquired, distributed, retained, and assessed warrants intense study and interest.

2. SOCIAL EPISTEMOLOGY AND EPISTEMIC AGENCY

Practitioners of social epistemology, despite all the divergent ways they conceive it, tend to have some approaches in common. In particular, there are arguably two main avenues social epistemologists have taken. The *first* avenue explores the *contributions* social influences make on the acquisition of knowledge. This can be loosely divided into the following, in which some or all of these topics may be investigated: 1) the social mechanisms that enable "justified true belief" (i.e., the traditional view of knowledge), 2) the social mechanisms that potentially undermine traditional views of knowledge, and 3) the manner in which norms (i.e., social expectations and conventions) affect the conception of knowledge and epistemic practices. The *second* avenue concerns the *evaluation* of the social impact and moral outcome of the acquisition, assessment, dissemination, and retention of a particular epistemic view or practice. Hence, this avenue may address any of the above themes from a moral or political perspective.

Insofar as persons are responsible for their actions, and may be responsible for having some reasonable expectations of what their unforced actions may engender, one is an "agent." It is one's status as an agent that renders one responsible for one's actions. Epistemic agency is central to social epistemology in two ways. In the *first* instance, if the acquisition of knowledge requires significant choices, such epistemic decisions can make the agent accountable according to how well or how poorly their decisions were formed. In the *second* instance, there is, as already noted, a moral component to epistemic agency. For example, *if* it is the case that knowledge (or much of it) is irreducibly a social phenomenon, in that individuals, agents, or groups (e.g., corporations) play significant roles in the acquisition, assessment, dissemination, and retention of knowledge, *then* one can evaluate epistemic agents in respect to the ethical, political, and practical implications of their epistemic activities and practices.

Epistemic agency has enormous political implications. For instance, just like any activity central to the sustainability of a society, the handling or mishandling of knowledge can allow its members to flourish and be free, or it can harm them.[3] In some cases, it can even limit freedom by indenturing individuals to those who possess knowledge. Indeed, it might intentionally be kept from a populace in order to render it ignorant of viable alternatives, as seen in oppressive regimes. In these and other respects, epistemic agency is centrally tied to moral and political aspects of social life, just as a vibrant economy is directly bound to the freedoms and opportunities of a nation. Put differently, when a population's potential is shackled by subsistence living, in which the aspirations for arts, politics, and philosophy are not viable pursuits, or when plans for one's own future are made inactionable due to extreme poverty, one's freedoms are limited and one's choices are restricted. All this is true of extreme ignorance, which limits the capacity to act freely in a deliberate manner. One, for instance, cannot make purposeful choices about the future without knowing what options are viable, risky, or impossible. The type of knowledge that is available to a public can thus shape and orient its capacity to create opportunities, become morally aware, and participate in the political process. Framed in this context, concerns over what *can* be known, become overshadowed by the question "given that different types of knowledge permit different types of moral and political action, what *should* we know?" Epistemic agents are, to varying extents, responsible for their contribution (or lack of contribution) to how their community answers this question.

3. KNOWLEDGE: A COLLECTIVE ENDEAVOR

A core feature of this book is the claim that epistemic agency, like the topic of knowledge itself, should not be conceived as purely an individualistic endeavor. This theme is indicated in the title of this text, which refers to "decentralizing" the epistemic agent. In short, if we take epistemic responsibility to not rest solely in the individual who seeks knowledge but also in the entire epistemic community in which a potential knower is a member, we have a *diffused*, rather than a singular center in which epistemic agency lies. For instance, there is a sense in which the contemporary scientist is never solely responsible for his or her discovery, if by this we mean the following: that scientific discovery rests on 1) collaboration with one's peers (e.g., research groups, sharing of data, etc.), 2) discoveries made by one's predecessors, and 3) expected practices, methods, models, etc., that are regulated by a particular scientific community (such as randomized trials, blind review, strict control over testing environments, etc.). If scientists break too radically from the conventions of their predecessors or peers without seeking to amend them from within the system, their views will likely be rejected and their claims will unlikely become a part of public awareness. Additionally, there are innumerable supporting roles of technicians, administrators, financial backers, educational institutes, search engines, libraries, etc., that indirectly facilitate and influence the epistemic flavor of scientific knowledge. Equally so, knowledge in any *field* does not lie solely in individuals but rather in a host of contributing factors that exceed the activities of individual actors.[4]

For the above reasons, the epistemic agent should *not* be considered an island unto itself (as traditional epistemology portrays the human subject). Instead, the contributing authors of this text argue that the epistemic agent is a participant in knowledge production, or in some cases, a contributing member of corporate bodies or nonhuman assemblages, which shape our current status as knowers. Here then, a preliminary definition of "epistemic agency" might be broadly construed as anything that is *responsible* for the assessments, acquisition, dissemination, and retention of knowledge. However, the issue as to whether the term *epistemic agent* should reference individuals, social groups, nonhuman entities such as corporations, complex computers, and/or all of the above, is an important subtheme of epistemic agency that echoes throughout this collection, making it a resource for approaching competing views on how social epistemology and epistemic agency *ought* to be conceived.

A final note of clarification before proceeding: the term *epistemic agency* is sometimes used as a technical term concerning whether or not and individual can *choose* the beliefs that underlie epistemic claims.[5] This text assumes a broader definition of the term. As human beings, social groups, and corporate entities can be responsible for influencing all aspects of morally relevant behavior, so too can individuals, social groups, and corporate entities be responsible (or held accountable) for how they influence all aspects of knowledge. The authors of this text address epistemic agency only in this broader context.

3. THREE CHALLENGES FACING SOCIAL EPISTEMOLOGY

In order for social epistemology to hold value as a subdiscipline of philosophy, it must be able to overcome three obstacles. In discussing these obstacles, I will further develop what it means to be engaged in social epistemology, which, like most fields of study, offers numerous entry points and areas of disagreement.

One, social epistemology must be able to offer significant arguments against epistemic approaches that assume the individual is *solely* responsible for what he or she knows or fails to know. Mainstream epistemology has important roots in Descartes. Descartes believed, like most contemporary philosophers, that there is one true way the world is, independent of how humans may think about it. He claimed that the individual possesses mental representations of what the world is like, and these representations can either be accurate or inaccurate. If they are accurate, they are considered true. If they are inaccurate, they are considered false. This briefly outlines what has been called the correspondence theory of truth: in order for something to be true, one's mental representation needs to coincide with existence. What made Descartes and his intellectual descendants individualistic is the following belief: since mental representation can only occur in an individual's head, and since these representations are true or false according to how the individual conceives and models them, the individual is *solely* responsible for his or her status as a knower. If this type of individualistic conception of truth tells the *whole* story of knowledge, there is little to no need for a subdiscipline called "social epistemology," as the social dimension of knowledge would be of little importance. Social epistemologists thus need to offer socially centered alternatives to this view of knowledge or find ways in which social participation plays an important role in our representations of existence.

Two, social epistemology must do more than merely offer a description of the conditions in which a particular culture endorses knowledge claims, which Sanford Goldberg calls *Descriptivism about Normativity* (see chapter 1 of this text). Frederick F. Schmitt offers a parallel insight:

> Social epistemology is the conceptual and normative study of the relevance of social relations, roles, interests, and institutions to knowledge. Thus it differs from the sociology of knowledge, which is an empirical study of the contingent social conditions or causes of knowledge or of what passes for knowledge in a society, a study initiated by Karl Mannheim. (1).[6]

Social epistemology must be able to offer epistemic insights via *argumentation* that empirical study cannot provide or else it would be incapable of making positive contributions that are qualitatively distinct from "sociology of knowledge."[7] One difference is that social epistemology aspires to be normative, and not just descriptive. From an empirical study of what is, one cannot arrive from that alone at what *ought* to be. If social epistemology cannot offer some form of "ought" regarding knowledge, its assessment, acquisition, or dissemination, at either a moral or epistemic level, then there is no reason to consider it a worthy philosophical approach. For if this were the case, one could not even argue that one form of justification *should* be sought over that of another.

Three, insofar as social epistemology is concerned with numerous and often opposing social influences that contribute to the acquisition, dissemination, and assessment of knowledge, it must avoid *"unrestrained relativism."* Unrestrained relativism offers no basis for evaluating knowledge claims and therefore presumes that all epistemic claims are of equal value. Failure to avoid this formulation of relativism would render philosophical arguments for or against epistemic claims pointless. Hence, without normative commitments that enable evaluation of epistemic claims, social epistemology could only describe epistemic claims (as opposed to argue for or against them or critique value claims). Such circumstances would destroy any plausible distinction between social epistemology and sociology of knowledge (as described above) in that it would merely be a descriptive project.

If all social epistemologists were explicitly confronted with the above three challenges, there would be very little consensus. Take for instance the two most popular camps working on social epistemology. In one camp, one finds Alvin Goldman and his followers, who support traditional individual-focused conceptions of truth. This seems to be the case for most analytic philosophers working in this field. For instance, Goldberg's Anchor Chapter (in this text) accepts the fundamentals of traditional individualistic formulations of truth—namely that knowledge, despite significant social contributions, irreducibly concerns objective states of existence that are, in theory, distinguishable from social influence, changing cognitive outlooks, and linguistic paradigms. Such conceptions can foster the idea that the social dimension of knowledge is superfluous once knowledge of existence is obtained. In this regard, the social contributions to knowledge are nothing more than a vehicle to obtain knowledge and that such a vehicle should be distinguished from "objective truth," as Goldman claims.[8]

Those in Steve Fuller's camp (and those who do not side with Fuller, but also reject individual-based notions of truth) tend to see knowledge as being socially constructed. As such, they tend to see knowledge as being subject to historical change, which leads to certain relativistic views concerning how knowledge *ought* to be evaluated. Such views also tend to be "constructivistic," in that standards of what it *means* to know and how knowledge is *deemed* to be achieved are considered social creations.

Social epistemologists can thus be seen as running the full gambit from accepting traditional views of truth but chipping away at their purely individualistic tendencies, to whole scale rejection of the traditional model of truth. Those in the latter camp advocate for a loosely restrained relativism that permits *some* moderate forms of epistemic assessment.[9] In this latter case, the assessment tends not to concern the critique of knowledge qua knowledge but rather its moral and political implications.

4. THE STRUCTURE OF THIS TEXT AND A SYNOPSIS OF ITS CHAPTERS

This text is unique in that, not only is it the first book-length text to focus primarily on epistemic agency, it is also atypical in that it centers on Anchor Articles (written solely for this text) by influential thinkers in the field of social epistemology—Sanford C. Goldberg and Steve Fuller. These two authors represent rather divergent ways of approaching social epistemology. Goldberg, for instance, works in the analytic tradition and accepts many of the traditional views of knowledge, whereas Fuller vehemently opposes the analytic style and traditional models of truth. Each subsequent author takes a stance on some aspect of one or both of these anchor chapters. Goldberg's and Fuller's contributions thus "anchor" the entire text by ensuring shared continuity of content.

In the first Anchor Article, "A Proposed Research Program for Social Epistemology," Goldberg argues that the research program for social epistemology ought to be "the systematic exploration of the epistemic significance of other minds." He develops this program by asking the following questions:

1. What are the varieties of ways we rely on others in information-acquisition, -storage, -processing, -assessment, and -transmission?
2. With respect to each of these sorts of ways we rely on others, what is the epistemic significance of the fact that we do so rely?

He concludes by arguing that traditional epistemology can and should be a core contributor to this sort of inquiry but that it "should provide a distinctly normative framework with which to supplement empirically-minded descriptive answers" to the first question. In so doing, he argues, "it can suggest precisely where questions of epistemic agency bear on questions of epistemic assessment."

In the second Anchor Article, "A Sense of Epistemic Agency Fit for Social Epistemology," Fuller begins by writing that his "vision of social epistemology has been always committed to a strong sense of 'normativity,' understood in a sense familiar from legal and political theory, yet seemingly alien to the thinking of analytic social epistemologists." Fuller's version of social epistemology sees both epistemic claims and epistemic agency as largely constructed by changing and dynamic social factors, which are best analyzed and assessed by principles common to legal and political theory.

After the Anchor Articles, the topic of analytic social epistemology and its alternatives are addressed. Finn Collin, in "Two Kinds of Social Epistemology and the Foundations of Epistemic Agency," examines the manner in which Goldberg and Fuller can be interpreted as both contributing to the same epistemic project, albeit from different sides of a spectrum. He also examines the extent to which their notions of "mindedness" (consciousness) in social epistemology and the problem of non-individual epistemic agents require further explication.

Francis Remedios and Valentine Dusek in "Fuller's Social Epistemology and Epistemic Agency" argue that Fuller's account of epistemic agency is more effective than traditional analytic social epistemological approaches to epistemic agency. They argue that its efficacy lies in its inclusive and malleable conception of agency that takes into account a host of changing conditions, such as developments in political and corporate entities, human enhancements, technology, and science. Despite their defense of Fuller's account of agency, they close by arguing that Fuller's account of scientists and their agential activates is too limited.

The question as to whether there are limits to epistemic agency is addressed by Paul Faulkner and Fred D'Agostino. Faulkner argues in "Agency and Disagreement" that contrary supporting evidence provided by experts and epistemic peers illuminates the manner in which epistemic agents are forced to choose among different and often equally supportable means of justification. He further argues that epistemic choices are not only most apparent during active disagreements, but that the lack of disagreement can lead to circumstances in which there is little opportunity for epistemic agency. D'Agostino's chapter "Disciplines, the Division of Epistemic Labor, and Agency" offers an analysis of the different forms of epistemic agency afforded by traditional fields of inquiry and considers the motivating factors for engaging in their disciplinary practices. He concludes that many of the motivating factors for participating in a particular discipline, along with the specific epistemic roles they afford, often restrict epistemic choices.

Though the importance of human and nonhuman epistemic agency is acknowledged by nearly all the contributing authors of this text, Orestis Palermos's and Duncan Pritchard's chapter "The Distribution of Epistemic Agency" and Frank Scalambrino's chapter "Toward Fluid Epistemic Agency: Differentiating the Terms Being, Subject, Agent, Person, and Self" directly grabble with this topic. Palermos and Pritchard offer an accessible introduction to their view on extended knowing (i.e., the manner in which knowledge extends beyond the minds of individual knowers) and requires what they call "cognitive integration." They also challenge Goldberg in arguing that epistemic agency extends beyond individual knowers. Scalambrino's chapter argues that human epistemic agents can be existentially threatened and oppressed by nonepistemic agents, such as corporations. In particular, he argues that nonhuman epistemic agents can limit or oppress the existential horizon in which human agents act and think. He further argues that it is this difference that should distinguish human epistemic agency from that of nonhuman epistemic agents.

The last section of this book focuses on social epistemology and the contributions German idealism can make to it. Angelica Nuzzo, in "'Epistemic Agency': A Hegelian Perspective," argues that Goldberg and Fuller provide ambiguous and redundant notions of epistemic agency. She asks: "what is the *specific* issue that the notion of 'epistemic agency' intends to distinctively address beyond these quite obvious ambiguities and redundancies, i.e., beyond the minimal notion of agency implied by some epistemic attitudes, and beyond the minimal notion of knowledge implied by all types of practical activity?" Her answer is one that borrows heavily form the Hegelian tradition, in which the agent begins as an individual who comes to understand his/her place in the world and becomes socially oriented through taking action. This development of the individual is centered on activities that define the agent, as opposed to some intrinsic state that defines a person. In line with Hegel, she argues that the mature agent overcomes the seeming dualities of the individual and social by fully integrating both components into her consciousness.

Patrick J. Reider in "Epistemic Agency as a Social Achievement: Rorty, Putnam, and Neo-German Idealism" argues that epistemic agency (for human beings) is irreducibly a social enterprise that shares the same social-linguistic preconditions as knowledge. He analyzes Richard Rorty's rejection of both realism and idealism in favor of pragmatism that assesses epistemic claims by their social usefulness and the demands of causal restraint. While Reider argues that Rorty is correct to reject the realist epistemic project, he also argues that his case against idealism fails. Reider argues for a variation of neo-German idealism that illuminates the social-linguistic nature, preconditions, scope, and limits of epistemic agency.

NOTES

1. I am grateful to John Lyne for editorial comments on an earlier draft of this text.
2. I am not claiming that moral, religious, or aesthetic assessment require some known standard, which renders such judgments correct or incorrect. Rather, my point is that these types of assessments require knowledge of the conditions relevant to one's assessments. For instance, if one makes an assessment without any knowledge of what properly relates to the moral claim under consideration, then one's assessment arguably has no merit.
3. See Patrick J. Reider, "A 'Dialectical Moment': Desire and the Commodity of Knowledge," in *The Future of Social Epistemology: A Collective Vision*, ed. James Collier. (London: Rowman & Littlefield International, 2016).
4. Additionally, John Lyne has argued that rhetorical activity contours variable forms of communicative reasoning and provides the link between individual and public processes of knowing. See, e.g., John Lyne, "'Having A Whole Battery of Concepts': Thinking Rhetorically About the Norms of Reason," *Social Epistemology* (online special issue on "Normative Functionalism and the Pittsburgh School"), January, 2013; and "Rhetoric and the Third Culture: Scientists and Arguers and Critics," in Mark Porrovecchio, ed., *Reengaging the Prospect of Rhetoric* (Routledge: 2010), 132–52.
5. See Pascal Engel, "Is Epistemic Agency Possible?" *Philosophical Issues*, 23 (2013), 158–78.
6. Frederick F. Schmitt "Socializing Epistemology: An Introduction through Two Sample Issues," in *Socializing Epistemology: The Social Dimensions of Knowledge*, ed. by Frederick F. Schmitt (Lanham: Rowman & Littlefield Publishers, 1994), 1.
7. Despite this distinction, Fuller, Goldberg and many others, myself included, have maintained that social epistemology can and should be a multi-disciplinary pursuit.
8. Alvin I. Goldman, "Social Epistemology: Theory and Applications," *Royal Institute of Philosophy Supplement*, 64 (2009), 118.
9. For a broad overview of differing perspectives on social epistemology see James Collier's edited volume, *The Future of Social Epistemology: A Collective Vision* (London: Rowman & Littlefield International, 2016).

REFERENCES

Collier, James, editor. *The Future of Social Epistemology: A Collective Vision*. London: Rowman & Littlefield International, 2016.
Engel, Pascal. "Is Epistemic Agency Possible?" *Philosophical Issues*, 23 (2013), 158–78.

Goldman, Alvin I., "Social Epistemology: Theory and Applications," *Royal Institute of Philosophy Supplement*, 64 (2009), 1–18.

Reider, Patrick J., "A 'Dialectical Moment': Desire and the Commodity of Knowledge," In *The Future of Social Epistemology: A Collective Vision*. edited by James Collier. (London: Rowman & Littlefield International, 2016).

Schmitt, Frederick F. "Socializing Epistemology: An Introduction through Two Sample Issues," In *Socializing Epistemology: The Social Dimensions of Knowledge*, edited by Frederick F. Schmitt. USA: Rowman & Littlefield Publishers, 1994.

I

Anchor Articles

ONE

A Proposed Research Program for Social Epistemology

Sanford C. Goldberg

In the last thirty years or so, philosophers, social scientists, and others have begun to speak of and pursue inquiry within a distinctly "social" epistemological framework.[1] In one sense, the fact that there should be a "social" epistemology is easy to explain. As standardly conceived, epistemology is the theory of knowledge. As standardly practiced, the theory of knowledge is interested in the various *sources* of knowledge. We might think to explain the existence of a distinctly "social" epistemology, then, in terms of the existence of distinctly social sources of knowledge.

While there is much to this explanation, it is at best incomplete. For one thing, our community is implicated in our body of knowledge in ways that go far beyond that of being a source of information. For another, this purported explanation fails to make clear precisely why many theorists think that social epistemology presents a challenge to certain aspects of the epistemological tradition. In order to appreciate the nature of this challenge and to deepen our explanation for the existence (and rationale) of a distinctly social epistemology, we would do well to revisit the reasons for thinking that there are distinctly social sources of knowledge. These reasons point to a central deficiency in traditional epistemology. Once we recognize this deficiency, we will be in a position to appreciate the variety of different ways in which knowledge acquisition is (often, and perhaps even typically) a social activity.[2]

Traditional epistemology is individualistic in its orientation: it focuses on the states, skills, and background information of individual epistemic subjects. As such it recognizes only two general ways for an individual to acquire knowledge of her environment: through perception (broadly construed), or through inference (relying on one's own background information). Once this framework is accepted, we are limited in the role(s) that other people can be recognized as playing in one's pursuit of knowledge. Put in the starkest terms possible, others' antics and appearances—their doings and sayings, their dress and manner of presentation etc.—are no different in principle from the antics and appearances of *any* of the objects in one's environment. That is to say, on this individualistic framework other people's antics and appearances have the status of *evidence* from which one can come to know things through inference. On this picture, when I come to know, for example, that the dean is in London through your telling me that she is, my route to knowledge here is no different in kind than the route by which I come to know that it's cold outside by seeing my brother reach for his parka, or the route by which I come to know that we have a mouse problem by observing the mouse droppings under the sink, or the route by which I come to know that it is currently raining by hearing the characteristic patter-patter-patter on my roof. In each case perception makes available to me a piece of evidence—an utterance of yours; a piece of nonverbal behavior of my brother's; mouse droppings; sounds coming from the roof—from which I go on to make inferences. In making these inferences, I am relying on my background information, both for interpreting the evidence in the first place (you have *asserted that the dean is in London*; those things are *mouse droppings*; that's the sound of *rain*), and for knowing which inferences to draw from my evidence (e.g. *your asserting something is highly correlated with the truth of what you've asserted; the presence of mouse droppings is highly correlated with the nearby presence of mice; my brother's reaching for his parka is highly correlated with it's being cold outside*).

I believe that this approach to the role others play in one's pursuit of knowledge fundamentally mischaracterizes that role. To be sure, we often do draw inferences from others' antics, speech, and appearances; and when we do, their antics, speech, and appearances serve as evidence. But there are also cases in which we rely on others *as epistemic subjects in their own right*. Perhaps the clearest and most straightforward example of this—but by no means the only one—is the case of testimony. When we accept another's word for something, we regard them not merely as providing potential evidence, but also, and more centrally, as manifesting *the very results of their own epistemic sensibility*. When one comes to acquire knowledge in this way, it is plausible to think that the epistemic task has been socially distributed; and we might speculate that subjects who share knowledge in this way constitute (part of) an epistemic community. That is to say, they are members of a group whose knowledge environment is structured by various social practices regarding the acquisition, storage, processing, transmission, and assessment of information.

I cannot pretend that this picture is anything but controversial among those who grew up in the more traditional, individualistic orientation that characterizes orthodox epistemology. Even so, I won't defend this picture further here.[3] Instead, I would like to suggest how social epistemology looks from the vantage point of those who take this picture seriously. For those who do take this picture seriously, we stand in a fundamentally different relation to other epistemic subjects than we do to the rest of the items in our environment. Since the point at issue reflects the roles epistemic subjects play as epistemic agents in a common epistemic community, it will be helpful to begin by saying a few words about epistemic subjects, epistemic agents, and epistemic communities.

Throughout this chapter I will want to be able to refer to the sort of entity of whom we can intelligibly ascribe knowledge and other epistemic states (such as justified or rational belief). I will use the term *epistemic subject* to do so. Thus other people are epistemic subjects: we can intelligibly affirm or deny that Smith knows that it is raining, or that Jones believes with justification that the economy will improve. But so too some collectives might be epistemic subjects as well: at any rate we do affirm or deny such things as that the Obama administration knows that immigration laws in the United States need to be addressed, or that the firm's engineering team believes with justification that the bridge is no longer structurally sound.[4] In addition to speaking of epistemic subjects, I will sometimes want to highlight the various roles that epistemic subjects play in acquiring, storing, processing, transmitting, or assessing information. When I want to highlight these roles, I will speak of them (not as epistemic subjects, which they remain, but rather) as *epistemic agents*. This difference in nomenclature—between "epistemic subject" and "epistemic agent"—marks a *notional* difference: to speak of an epistemic agent is to speak of an epistemic subject, albeit in a way that highlights the role(s) played by the subject in the process(es) by which knowledge is acquired, stored, processed, transmitted, or assessed. Finally, I will also be speaking of the practices, institutions, and norms that structure the relations between epistemic agents as they go about their information-seeking business (both individually and socially); to do so I will speak of their shared "epistemic community."

Having introduced these terms, I can now proceed to describe in more detail how (from the epistemic point of view) our relations to other epistemic subjects differ from our relations to the rest of the items in our environment. Here I highlight three dimensions of difference. These dimensions correspond to what I will proceed to call the core project of social epistemology: that of characterizing the *epistemic significance of other minds*.

The first way in which our relations to other epistemic subjects differ from our relations to the rest of the items in our environment is this: epistemic subjects stand in various *epistemic dependency relations* to other epistemic subjects in their shared epistemic community. The basic idea of an epistemic dependency relation can be brought out in terms of the nature of epistemic assessment itself, in which we assess a subject's belief (or her degree of belief) in a proposition. Such assessment aims to characterize how well-supported her belief is (alternatively: whether that degree of belief is warranted by her evidence). This sort of assessment is a fully normative affair, since it appeals to standards (e.g. of rationality, epistemic responsibility, and/or reliability, among other standards) whose satisfaction is required if the subject's (degree of) belief is to count as amounting to justified belief or knowledge. I describe one subject (S_2) as *epistemically dependent* on another subject (S_1), then, when an epistemic assessment of S_2's belief—an assessment along one or more of the dimensions just described—requires an epistemic assessment of the role S_1 played in the process through which S_2 acquired (or sustained) the belief. (As we will see below, it will be helpful to think of S_1 here not merely as an epistemic subject but as an epistemic agent.) It is of course a substantial assumption that we do exhibit epistemic dependence on others; traditional epistemology would deny this. But social epistemology as I understand it embraces this assumption, and with it recognizes that our epistemic tasks are often socially distributed among the members of our epistemic community. Relying on another person's say-so is *one* kind of epistemic dependence (for which see Goldberg 2010); but it is not the only kind, and it is a task of social epistemology to enumerate and describe the variety of kinds of epistemic dependence exhibited in our interactions with others.[5]

A second way in which our relations to other epistemic subjects differ from our relations to the rest of the items in our environment lies in the variety of norms that enable us to calibrate our expectations of one another as epistemic agents, as we pursue our inquiries (whether individually or jointly). Consider for example the expectations you have when you rely on your doctor, or your lawyer, or your accountant. You expect them to be knowledgeable in certain ways, to be apprised of the best practices, to be responsive to any relevant developments in their specialties, and so forth. Alternatively, consider the expectations you have of the other members of your research team, or of your business partners. You expect them to do their jobs properly, to notify the rest of the team (the other partners) if there are any developments that bear on the research of the whole team (the success of the business), and so forth. Or consider the expectations you have of your neighbors, friends, and family members. When you have long and mutually acknowledged traditions of informing one another of the news in certain domains, you come to expect this of one another. Or, to take a final example, consider the expectations you have when you encounter someone who tells you something. You expect her to be relevantly authoritative regarding the truth of what she said. In many and perhaps even all of these cases, the expectations themselves reflect various norms that regulate our interactions with other epistemic agents. In some cases, these norms are provided by professional or institutional organizations, and rationalize our reliance on members of those professions or institutions; in other cases, the norms in question are established explicitly, as a matter of agreement, e.g. among team members or business partners; in still other cases, the norms themselves are part of the practices (e.g. of information-sharing) that emerge over the course of repeated interaction between the parties, after the parties mutually (if perhaps only implicitly) acknowledge their mutual reliance on certain aspects of the practice; and in still other cases, the norms are part of sophisticated social practices (such as those regarding the practice of assertion[6]) whose features are, if only implicitly, mutually acknowledged by all participants. (This is not intended to exhaust the possibilities.)

Norm-sanctioned expectations, I submit, are not so much *predictions* of the behavior of our fellows—although they may give rise to such predictions—as they are *normative expectations* of our fellows. For example, your expectation that your doctor knows best practices for the treatment of your condition is not (or not merely) based on the evidence that doctors *are* generally reliable in this way; rather, it constitutes something you normatively expect of her. It is akin to parents' expectation that their teenager will be home by midnight (an expectation to which they are entitled even if their teenager has a long history of staying out too late). These expectations enable us to solve complicated coordination problems we face as we seek to acquire knowledge in communities that exhibit a highly differentiated division of intellectual labor. I regard it as a central task for social epistemology to enumerate and describe the norms that underwrite these expectations, to articulate their epistemic bearing on the *predictive* expectations they underwrite, and ultimately to evaluate the norms themselves in terms of their role in securing true belief and knowledge.[7]

Since the notions of normative and predictive expectations will loom large in the sections to follow, it is important to be clear about the relationship between them. To a first approximation, one epistemic agent, S_2, *normatively* expects something from another epistemic agent, S_1, when S_2 holds S_1 *responsible* in the relevant way. Such normative expectations are warranted by the norms of prevailing practice (for a defense of which see Goldberg (forthcoming)). When a normative expectation is warranted in this way, I will speak of agents' *entitlement* to have the normative expectation in question. As I noted above, there are two fundamental theoretical questions regarding normative expectations. First, given a set of normative expectations sanctioned by the norms of a given practice, do these expectations actually conduce to epistemically good outcomes? In asking this, we are taking a critical perspective on the norms and practices of a given community, with the aim of assessing how well these norms and practices serve epistemological ends. (As I will argue in section 3, this is one place where the traditional normative vocabulary of e.g., epistemology, will come in handy.) Second, how do the normative expectations to which an agent is entitled relate to corresponding *predictive* expectations she has? The latter expectations are a species of belief (about the future), and hence are straightforwardly assessable from an epistemic point of view. But it remains to be seen how being entitled to hold someone responsible for an outcome relates to the justification one has for *believing* that one will get that outcome. And this point brings me to the third way in which our relations to other epistemic subjects differ from our relations to the rest of the items in our environment.

Given the normative expectations one has on those on whom one epistemically relies, as well as the epistemic dependence that results thereby, the epistemic assessment of beliefs formed through one of the "social routes" to knowledge (Goldman 2002; see also Goldman 1999 and 2009) is decidedly different from the epistemic assessment of beliefs not so formed. Insofar as epistemic tasks really are socially distributed, our assessment itself must be a social one. It must take into account not only the other individual(s) on whom the belief epistemically depends, but also the social practices and the norms that regulate these practices. This will include the various practices and norms that constitute what we might call the "epistemic environment" in which agents go about their knowledge-seeking business—and (as noted in the preceding paragraph) their relationship to the justification of belief. To the extent that one's epistemic environment bears on the proper assessment of one's beliefs, we will need to rethink the nature of epistemic assessment, in a way that reflects the various epistemic dependencies and social norms that are implicated in the production and sustainment of belief. I regard it as a task for social epistemology to reconceive the nature of epistemic assessment, and, where needed, to reconceive the categories employed in the assessment.[8]

In short, what I would call the *epistemic significance of other minds* can be seen in (i) the various forms taken by our epistemic dependence on others, (ii) the variety of norms that underwrite our expectations of one another as we make our way in the common epistemic environment, and (iii) the distinctive epistemic assessment(s) implicated whenever a doxastic state is the result of a "social route" to knowledge. In characterizing (i)–(iii) we aim to capture the ways in which our relations to other epistemic agents differs from our relations to the rest of the items in our environment. And it is this, I propose, that provides the rationale for a distinctly social epistemology: social epistemology ought to be the systematic investigation into the epistemic significance of other minds, where this is understood to involve the epistemic tasks I have described in connection with each of (i)–(iii).

There are several lessons to be drawn from the foregoing rationale.

First, if the foregoing rationale is to be our guide, it is an open question to what extent epistemology is social. To be sure, it is social at least in part because we depend on others as sources of knowledge. But this does not exhaust the roles others play in our pursuit of knowledge. Consider the roles other play as experts, as well as in the certification of expertise; in policing standards of assertoric speech and writing; in peer review; in professional organizations, where there is a need to articulate and police standards of professional (including intellectual) behavior of its members; in devising technologies aimed at enabling us to discern more of nature's secrets, and in training others how to use that technology; and in the process by which we educate our young to become thoughtful, critical, productive members of our own knowledge community. In all of these (and no doubt many other) ways, we depend on each other epistemically. As I see matters, it is the task of social epistemology to enumerate and describe these ways, to characterize the norms that underwrite our expectations of one another in these efforts, and to evaluate these norms in terms of their role in securing true belief and knowledge. None of us should be confident of the precise contours of social epistemology (or its place in epistemology more generally) in advance of an extended investigation into these matters.

A second lesson is this: it is an open question how best to understand the role of technology in inquiry. Here I mean to include not only tools of communication and information technology but also the distinctive technology and instrumentation employed in mathematics and the social, natural, and human sciences. On one hand, technology falls within that part of the world whose antics, it would seem, provide us merely with evidence. On the other, technology itself is typically the result of a good deal of epistemic effort: other epistemic subjects bring their epistemic sensibility to bear on the construction, validation, employment, and teaching of technology. What is more, at least some of our technology—here I have in mind scientific instruments—is specifically designed for the purpose of providing results which represent, or at least enable us to discern, aspects of the world's features. Thus our reliance on technology does appear, in both indirect and direct ways, to involve reliance on other epistemic agents. It would thus seem that social epistemology would do well to explore the epistemological dimension of our reliance on technology.

And there is a third lesson as well: it is an open question whether the solitary epistemic subject is the only proper unit of analysis at which to conduct epistemic assessment. So far I have been speaking as if the unit of analysis *is* the individual subject. But many social epistemologists will take issue with this assumption. The development and evaluation of the case for and against this assumption ought to be on the agenda of social epistemology.[9]

All three of these lessons suggest that we do a great disservice to the potential of social epistemology research if the only rationale we recognize for pursuing this research is that there are social sources of knowledge. Such a conception is far too limited. In studying the variety of epistemic dependency relations, the set of epistemic norms that enable epistemic tasks to be socially distributed, and the nature of the epistemic standards used to assess the resulting beliefs, we can see clearly why social epistemology is not merely one category among others on the list of sources of knowledge. We can also see the sort of challenge that social epistemology presents to orthodoxy. In a nutshell, our epistemic dependence on others cannot be understood in the orthodox (individualistic) terms of traditional epistemology, nor can the questions about the nature and scope of this dependence find their place among the standard questions of individualistic epistemology. It would thus be a significant mistake to think that acknowledging the relevance of social epistemology is merely adding one other item to the list of knowledge sources.

In short, I submit that the pursuit of social epistemology is the attempt to come to terms with the epistemic significance of other minds. There is a straightforward rationale for making such an attempt: other people are (not mere sources of knowledge, but) epistemic subjects in their own right who, through their epistemic agency, bring their own epistemic sensibility to bear in all sorts of ways as we shape and operate within a common epistemic environment.

1. SOCIAL EPISTEMOLOGY AND EPISTEMIC AGENCY

The foregoing provides a clear sense of the relevance of epistemic agency to issues of social epistemology. The link is provided by the variety of norms that underwrite our normative expectations of one another as we make our way in the common epistemic environment.

As I already mentioned in an endnote (note 6), one sort of norm that underwrites our normative expectations of one another as inquirers is what philosophers of language and epistemologists have called the "norm of assertion." Consider the sort of expectation you form when another person tells you something: if a person tells you that things are thus and so, you hold the speaker responsible for being in a suitably authoritative epistemic position to judge that things are thus and so. Precisely what this involves—whether this requires that the speaker know, or merely that she have good reason to believe—is a matter over which there is some debate. But the basic idea, that each of us expects others to be relevantly authoritative when they state that things are thus and so, is easily seen in our practices. (Consider how you would react on learning that what you were "told" was something the speaker had no good reason to think was true.)

Now it is a tricky matter to say precisely how the norm in question, requiring suitable authoritativeness when one makes an assertion, bears on the epistemic standing of beliefs that are acquired through accepting another's assertion. Does the presence of this norm, together with a hearer's absence of reasons to doubt the speaker's assertion, justify acceptance of that statement? Or does the hearer need to have additional positive reasons to think that the speaker lived up to the norm on this occasion?[10] But whatever one thinks about this matter, it seems patent that hearers do expect speakers to recognize that when an assertion is made the speaker renders herself answerable to the relevant expectation itself. And herein we see one dimension of agency in our social epistemic practices: speakers ought to act so as to conform to the standards that govern proper assertion.

Nor is the "norm of assertion" the only norm that bears on our wider practices in which information is acquired, stored, processed, transmitted, or assessed. Also relevant here are moral norms. For example, consider how the moral norm enjoining us to help others in need can generate an obligation to respond to another's need *for information*. Take a case in which a good friend asks you something. If you know the answer and have no good reason to refrain from responding, you should tell her. If you don't tell her under these conditions and she were to find out that you knew, she would be warranted in saying, "You should have told me so!" This sort of case gives added significance to the agency involved in satisfying the norm of assertion: not only must we regulate our speech so that we don't assert something when we *fail* to be relevantly authoritative, there are also cases in which we are under moral pressure to assert when we *are* relevantly authoritative. Among other things, this will require that we be responsible in determining *whether* we know the answer to a question presently before us.

The satisfaction of other norms structuring our epistemic environments illuminates still other aspects of epistemic agency. (Here I must be very brief.) The professional must be in a position in which she has all of the relevant evidence—that is, the evidence properly expected of her. One who relies on a doctor, for example, is entitled to expect that the doctor's degree of expertise and knowledgeableness conforms to all of the relevant professional standards, and that she (the doctor) stays abreast of all of the relevant developments in her specialization (perhaps under the regulation of a specialist group, such as that for pediatricians or radiologists, etc.). These standards themselves are norms that structure our reliance on doctors: they enable the ordinary citizen to take systematic advantage of the medical expertise in her community at only a small epistemic cost[11] to herself. Once again, the norms articulate what we properly normatively expect of the relevant individuals *as epistemic agents*: we expect that these individuals have acquired the evidence properly expected of them, that they have the knowledge properly expected of them, and in general that they have behaved with the sort of epistemic responsibility properly expected of them. Similar things can be said of our reliance on other professionals as well.

The prevalence of other norms regulating our epistemic communities highlights still other aspects of epistemic agency. Consider the sorts of expectations we have when we rely on familiar kinds of devices and instruments: thermometers and other temperature gauges, clocks, the instrumentation in our cars, GPS navigation, and so forth. Although it is often a straightforward matter to learn how to "read" these devices, most of us do not have anything beyond the most rudimentary understanding of how the devices themselves work. And, while many of us have enough experience with them to have empirically grounded confidence in their reliability, it is by no means clear that one *needs* to have such experience in order to be in a position to acquire knowledge through reliance on the devices. To see this, consider a young child just learning to tell time; and suppose such a child learns, not by looking at real clocks, but by looking at images of clock-faces, where she is told by her teacher how to read time by the orientation of the clock's hour and minute hands. Once the child learns to "read time" in this way, it would seem that she is in a position to know what time it is by looking at a clock, *even if her teacher never once testified to the reliability of clocks* (and even if the child herself did not have other evidence with which to confirm their reliability). Here it is natural to think that once we are properly initiated into a clock-using community, one can take for granted that clocks tell proper time. This presumption is defeasible, of course, and one must remain sensitive to the possibility that a particular clock on which one is relying is not working properly (and so is not a reliable indication of the time).[12] But the point is that one need not *also* acquire additional positive evidence for thinking that the clocks around here are reliable; the norms governing our interactions with clocks would appear to entitle us to *presume* as much, with defeat of this presumption requiring positive reasons for doubt in a given case. Once again, these norms correspond to expectations it would be proper to have regarding a range of epistemic agents: those responsible for the production, maintenance, and training in the usage of the devices and technologies themselves. (No doubt, the expectations to which we are entitled will vary according to the sort of technology in question, the prevalence of its use, and perhaps other factors as well. Our use of clocks may be atypical in this regard, in that our use of more technical instruments may not ordinarily come with an entitlement to make similar assumptions about their reliability.)

These last two cases are instances of a general point I made above. In pointing to the norms that structure our epistemic environments — the norms that warrant various sorts of normative expectations we have regarding the objects and people in our environment — we are pointing to expectations we have of one another as epistemic agents. We operate in an environment whose norms entitle us to form certain expectations, and hence to solve certain coordination problems that arise, when we epistemically rely on others in various ways. To be sure, we can make efforts to become aware of the various norms that structure our epistemic environment; and we can bring ourselves to reflect self-consciously on the evidence we have for thinking that things in general (or this person or that device in particular) reliably conform(s) to the norms. I surmise that most mature humans do have a good deal of relevant evidence, and that we do on occasion self-consciously reflect in precisely this way. But I submit that we typically do so only when we suspect that the situation doesn't seem right: when the person speaking to us doesn't appear to be fully confident, or is evasive; when one's watch has been making strange sounds recently; when the thermometer reads 20 degrees F, yet we know that it is in the middle of a Chicago summer; and so forth. What is more, I submit that there is a rationale for the otherwise curious fact that we are entitled to have various normative expectations within our epistemic community: having norms regulating our epistemic environment is precisely what enables us to focus our energies as agents on acquiring the results which, supposing that all is working as the norms require, provide us with detailed, sophisticated, and useful knowledge of the world around us. In other words, the norms themselves are part of what enable the division of epistemic labor to be as far-reaching and as systematic as it is.

2. AN INTERDISCIPLINARY PROGRAMME FOR RESEARCH IN SOCIAL EPISTEMOLOGY

In this penultimate section I want to use the foregoing characterization of social epistemology and its connection to epistemic agency, in order to describe the (seriously interdisciplinary) nature of the social epistemology research programme. This research will require the work both of those social science disciplines that study the prevailing norms and practices of our epistemic communities—disciplines such as history, sociology, political science, education theory and practice, psychology (cognitive and social), and anthropology—but also disciplines that bring a more "normative" orientation to the discussion—philosophy, law, parts of psychology,[13] economics, and perhaps others as well.

As I have been presenting matters, social epistemology is the systematic study of the epistemic significance of other minds (as manifested in connection with (i)–(iii) above). The emerging picture, then, is this: Epistemic subjects are conceived of as subjects in a community of knowers. Each subject has her own on-board cognitive competences; and among these cognitive competences are some that pertain to her role as an epistemic agent who aims to exploit the high-quality information that is available in her social environment—whether in the form of testimony, or in some other way in which information is acquired, processed, stored, assessed, or transmitted in a social way. Part of what enables this process to be as systematic and far-reaching as it is, I have suggested, are the set of norms that structure our epistemic environments. As agents, epistemic subjects operate in contexts in which a good deal of their interactions with other agents and with the world are regulated by standards pertaining to the flow of information in those environments. These standards (or norms) enable agents to get a bigger bang for their epistemic efforts than they would get in the absence of any such norms: were each agent forced to confirm for herself the reliability of the various sources of information available to her in her environment, and were she forced as well to confirm various hypotheses regarding what sorts of information are available (and how frequently), she would have to expend a good deal of effort simply learning about the various aspects of the flow of information within her environment—effort that might be better (more productively) spent learning about her world more generally.

From this description it is clear that there is an important role to be played in social epistemology by empirically minded social science. We need to know precisely what our information-relying practices are, what the norms or standards of those practices are taken to be, how (if at all) these norms or standards are communicated to members, how (if at all) they are enforced, how information does in fact travel through the network, and so forth. This constitutes what we might call the *structural description* of our various and overlapping epistemic communities. Without a detailed characterization of this sort, we would be theorizing blindly. Researchers from such disciplines as sociology, history, political science, psychology (especially cognitive and social psychology), education theory and practice, and anthropology are well suited to exploring these issues, and to providing the needed description.

But it is important to have the input of what I would describe as more "normative" disciplines as well. By calling a discipline "normative" I mean that it does not merely rest with a description of what our norms or standards are (and what they are *taken to be*), but instead is interested as well in what our norms or standards *could* and *should* be. As noted above, any theorist who hopes to address such normative questions must start off with a clear structural description of what those practices and their norms are (or are taken to be). But our normative theorist can then ask whether these practices and norms are proper, whether they are as they ought to be. That is, the theorist in one of the "normative" disciplines will normatively assess the very norms of our current practices, with an eye towards seeing whether these practices (and the norms taken to govern them) live up to the highest (epistemic, moral, or political) standards we have (or ought to have).

The "normative" disciplines I have in mind here include philosophy, law, areas of psychology, and economics. The normativity of these disciplines is seen in that they employ or explore norms or standards of assessment which can be used, in turn, to assess the normative standards of our current practices. The law brings in legal norms: it assesses practices in terms of their *legality*, and it assesses laws in terms of their *constitutionality*. Philosophy brings in epistemic as well as moral and political norms: epistemic standards enable us to assess beliefs in terms of such things as *reasonableness, rationality, reliability,* and *evidential well-groundedness*; moral standards enable us to assess actions in terms of their *moral goodness* or *badness*, or their *moral permissibility*; and (normative) political standards enable us to assess institutions and practices in terms of their *justice* and *fairness*. Parts of psychology have a distinctly ameliorative bent,[14] bringing in norms of reliability to assess the effectiveness of individual and group behavior: it assesses the behaviors of individuals and groups in terms of the speed and ease with which these behaviors produce their results, as well as the likelihood that those results will be true (or at least accurate enough for practical purposes). Finally, economics might be thought to be normative insofar as it aims to capture an assessment of the *rationality* of practices and actions (understanding rationality in terms of expected utility).[15] Given a structural description of our actual practices and the norms taken to govern them, we can assess these practices, as well as the individuals who participate in them, in the normative terms drawn from these disciplines. Do these practices live up to the highest ideals articulated in the normative disciplines? (And if they do not, should we conclude that it is the practices which need to change, or rather that we need to revisit our highest ideals?)

At this point I can imagine theorists from the more empirically minded social sciences recoiling at the thought that we can and should bring normative theory to bear on our actual practices. One worry on this score might be based on the idea that it is not the business of theorists to revise our practices; the best that we can do is describe those practices. Another more fundamental worry on this score might be based on a doubt whether there even is a "normative" orientation we can have beyond that provided by the practices themselves. Those who deny that there is such an orientation will endorse something like what I will call *Descriptivism about Normativity* (or 'DN' for short):

> (DN) Once we describe the social practices through which it comes to pass that things are *taken* as knowledge, *count* as good evidence, *pass for* being a justified theory, are *certified* as authoritative, are *regarded as* a legitimate criticism, and so forth, we will thereby have said what needs to be said about the relevant norms and standards themselves.

From the perspective of those endorsing DN, attempts to attain a metanormative perspective on these very practices will be naïve at best, impossible at worst.[16]

But these worries are unfounded, and it is important to appreciate why. When it comes to the practices that structure our epistemic environment, mere description is not the end of the social epistemology story. First, the practices themselves (as well as the standards taken to govern them) can conflict, and when they do so we will want some way to address this conflict. Second, it can come to pass that, in certain local contexts, we want to criticize the extant practices or norms, and we will then be in need of some normative orientation within which to cast our criticisms. To these two points it might be said that we should study how *the communities themselves* resolve matters when their knowledge practices reach opposing verdicts, or how *they themselves* criticize their own practices and respond to such criticisms, etc. This would be in keeping with DN and its fully descriptive characterization of the normativity of knowledge. But this response is inadequate. While we would do well to study such things, we would also do well to aim to occupy a critical perspective even when addressing a community's responses to its own internally-generated criticisms and difficulties. To do otherwise is to accept without criticism the community's own standards (or at least their standards for criticizing their standards). In addition to being groundless, such an acceptance risks degenerating into a thoroughgoing form of subjectivism.

Let me provide one illustration of how we might use the language of the more "normative" disciplines to address the adequacy of the standards of our actual knowledge practices. I have in mind the kinds of expectations we bring to bear in assessing others' assertions—the sorts of expectations described above. There are good reasons to think that even the most progressive-minded among us brings all sorts of implicit biases to bear as we do so—biases that systematically disfavor women and members of underrepresented minorities. If we induced our standards from our actual practices, the standards themselves would be decidedly unfair. But to make precisely this sort of point, we would do well to appeal the normative standards of epistemology, ethics, and normative political philosophy. (Indeed, the groundbreaking work of Fricker (2007) aims to do just this.) It would be a shame if we surrender the possibility of normative critiques of this sort out of a prior commitment to DN.

3. CONCLUSION

In this chapter, I have tried to articulate a research programme for social epistemology, understood to be the systematic study of the epistemic significance of other minds. In my presentation, such a programme involves addressing at least three things: (i) the various forms taken by our epistemic dependence on others, (ii) the variety of norms that underwrite our expectations of one another as we make our way in the common epistemic environment, and (iii) the distinctive epistemic assessment(s) that are appropriate whenever a doxastic state is the result of a "social route" to knowledge. I noted that it is in connection with (ii) that we see the most straightforward link between social epistemology and epistemic agency: the norms of our knowledge communities enable us to enhance the epistemic effects of our efforts beyond what they would be if each of us had to confirm for ourselves the various features of our epistemic communities. Finally, I argued that the study I envisage will require a healthy dose of both empirically minded social science as well as the input of the "normative" orientations found in disciplines like philosophy, law, ameliorative psychology, and economics.

NOTES

1. With thanks to Matt Kopec and Patrick Reider, for extensive comments on an earlier version of this paper.
2. 'often' or 'typically': I want to remain neutral on the issue whether ordinary perceptual knowledge is social in any interesting sense. (Those who think it is often appeal to the social dimension brought in by one's public language in shaping one's perceptual capacities.)
3. But see Sanford Goldberg, *Relying on Others: An Essay in Epistemology* (Oxford: Oxford University Press, 2010), and Sanford Goldberg, "The Division of Epistemic Labour," *Episteme* 8 (2001).
4. I say they 'might' be epistemic subjects: there is some dispute whether this talk of collectives as epistemic subjects is necessary, or whether it can be translated into talking about individual people and their relations to one another. I am neutral on this question here.
5. I make some initial taxonomic distinctions in Goldberg, "The Division of Epistemic Labour."
6. This idea is prevalent in the literature on the so-called "norm of assertion." See e.g. Timothy Williamson, *Knowledge and Its Limits* (Oxford: Oxford University Press, 2000), Jennifer Lackey, "Norms of Assertion," *Noûs* 41 (2007), and the various papers in Jessica Brown and Herman Cappelen, eds., *Assertion: New Philosophical Essays* (Oxford: Oxford University Press, 2011). Arguably, this idea can be traced back to a "deontic scorekeeping" view of assertion developed by Robert Brandom, "Assertion," *Noûs* (1983), and Robert Brandom, *Making it Explicit* (Cambridge: Harvard University Press, 1994). However, Brandom's approach to assertion is explicitly distinguished from the approach favored by the "norm of assertion" crowd in John MacFarlane, "What is an Assertion?" in *Assertion: New Philosophical Essays*, ed. Jessica Brown et al. (Oxford: Oxford University Press, 2011). See also Sanford Goldberg, *Assertion: On the Philosophical Significance of Assertoric Speech* (Oxford: Oxford University Press, 2015) for my attempt to take this idea and develop it into a full theory of the speech act of assertion.
7. I offer a framework in terms of which to theorize about these norms in Goldberg (forthcoming).
8. See Goldberg, *Relying on Others*; Goldberg, "The Division of Epistemic Labour"; and Sanford Goldberg, "Should Have Known," *Synthese* (forthcoming) for various extended arguments to this effect, and attempts to develop this sort of framework.

9. See e.g., Deborah Tollefsen, "From Extended Mind to Collective Mind," *Cognitive Systems Research* 7 (2006); Deborah Tollefsen, "Groups as Rational Sources," *Collective Epistemology* 20 (2011); Philip Pettit and Christian List, *Group Agency: The Possibility, Design, and Status of Corporate Agents* (Oxford: Oxford University Press, 2011); Jennifer Lackey, "Group Knowledge Attributions," In *Knowledge Ascriptions*, ed. Jessica Brown et al. (Oxford: Oxford University Press, 2012); and Miranda Fricker, "Group Testimony? The Making of A Collective Good Informant?" *Philosophy and Phenomenological Research* (forthcoming).

10. For reasons I explore in Goldberg, *Assertion*, chapters 2 and 3, something like this is the heart of a spirited debate in the epistemology of testimony.

11. The "small epistemic cost" is that those who would rely on doctors must be properly sensitive to indications that they are in the presence of a doctor, and must remain sensitive as well to signs of incompetence or insincerity even when in the presence of a doctor. (This requires less effort and less expertise than what would be required to attain expertise in the medical subject-matter itself!) Of course, this reduction in *epistemic* cost comes at a *financial* cost to ordinary citizens; but that is another matter.

12. Or that the recent change to daylight savings time makes salient the possibility that one's clock is an hour off. (With thanks to Matt Kopec for raising this possibility in this connection.)

13. I have in mind those parts which study effective/defective individual and group epistemic behavior, where effectiveness is determined in terms of the reliability with which such behaviors eventuate in beliefs that are true.

14. See, for example, Stephen Stich, *The Fragmentation of Reason* (Cambridge: MIT Press, 1990); and Michael Bishop and J. D. Trout, *Epistemology and the Psychology of Human Judgment* (Oxford: Oxford University Press, 2005). The description "ameliorative psychology" is taken from Bishop and Trout.

15. I acknowledge that not all economists will be happy with this normative characterization of economics!

16. Arguably, a view in the neighborhood of DN is held by Richard Rorty, *Contingency, irony, and solidarity* (Cambridge University Press, 1989); Steven Fuller, "Social Epistemology: A Quarter-Century Itinerary," *Social Epistemology* 26 (2012); and others in the science studies tradition.

REFERENCES

Bishop, Michael, and Trout, J. D. *Epistemology and the Psychology of Human Judgment*. Oxford: Oxford University Press, 2005.

Brandom, Robert. "Assertion." Noûs (1983): 637–50.

Brandom, Robert. *Making It Explicit*. Cambridge: Harvard University Press, 1994.

Brown, Jessica and Cappelen, Herman. *Assertion: New Philosophical Essays*. Oxford: Oxford University Press, 2011.

Fricker, Elizabeth. "The Epistemology of Testimony." *Proceedings of the Aristotelian Society*, Supplemental Vol. 61 (1987): 57–83.

Fricker, Elizabeth. "Against Gullibility." In *Knowing from Words*, edited by Bimal K. Matilal and Arindam Chakrabarti, 125–61. Amsterdam: Kluwer Academic Publishers, 1994.

Fricker, Elizabeth. "Telling and Trusting: Reductionism and Anti-Reductionism in the Epistemology of Testimony." *Mind* 104 (1995): 393–411.

Fricker, Miranda. *Epistemic Injustice: Power and the Ethics of Knowing*. Oxford: Oxford University Press, 2007.

Fricker, Miranda. "Group Testimony? The Making of a Collective Good Informant?" *Philosophy and Phenomenological Research* (forthcoming).

Fuller, Steven. "Social Epistemology: A Quarter-Century Itinerary." *Social Epistemology* 26 (2012): 267–83.

Goldberg, Sanford C. *Relying on Others: An Essay in Epistemology*. Oxford: Oxford University Press, 2010.
Goldberg, Sanford C. "The Division of Epistemic Labour." *Episteme* 8 (2001): 112–25.
Goldberg, Sanford C. *Assertion: On the Philosophical Significance of Assertoric Speech*. Oxford: Oxford University Press, 2015.
Goldberg, Sanford C. "Should Have Known." *Synthese* (forthcoming).
Goldmn, Alvin. *Knowledge in a Social World*. Oxford: Oxford University Press, 1999.
Goldman, Alvin. *Pathways to Knowledge: Public and Private*. Oxford: Oxford University Press, 2002.
Goldmn, Alvin. "Social Epistemology: Theory and Applications." *Royal Institute of Philosophy Supplement* 64 (2009): 2–18.
Lacke, Jennifer. "It Takes Two to Tango: Beyond reductionism and Non-Reductionism in the Epistemology of Testimony." In *The Epistemology of Testimony*, edited by Jennifer Lackey and Ernest Sosa, 160–92. Oxford: Oxford University Press, 2006.
Lackey, Jennifer. "Norms of Assertion." Noûs 41 (2007): 594–626.
Lackey, Jennifer. *Learning From Words*. Oxford: Oxford University Press, 2008.
Lackey, Jennifer. "Group Knowledge Attributions." In *Knowledge Ascriptions*, edited by Jessica Brown and Mikkel Gerken, 243–69. Oxford: Oxford University Press, 2012.
MacFarlane, John. "What is an Assertion?" In *Assertion: New Philosophical Essays*, edited by Jessica Brown and Herman Cappelen, 79–96. Oxford: Oxford University Press, 2011.
Pettit, Philip and List, Christian. *Group Agency: The Possibility, Design, and Status of Corporate Agents*. Oxford: Oxford University Press, 2011.
Rorty, Richard. *Contingency, irony, and solidarity*. Cambridge: Cambridge University Press, 1989.
Stitch, Stephen. *The Fragmentation of Reason*. Cambridge: MIT Press, 1990.
Tollefsen, Deborah. "From Extended Mind to Collective Mind." *Cognitive Systems Research* 7 (2006): 140–50.
Tollefsen, Deborah. "Groups as Rational Sources." In *Collective Epistemology* 20 (2011): 2–11.
Williamson, Timothy. *Knowledge and Its Limits*. Oxford: Oxford University Press, 2000.

TWO

A Sense of Epistemic Agency Fit for Social Epistemology

Steve Fuller

My vision of social epistemology has been always committed to a strong sense of "normativity," understood in a sense familiar from legal and political theory, yet seemingly alien to the thinking of analytic social epistemologists.[1] For me a "normative" approach is one committed to organizing the means available to bring about or maintain some desirable state of affairs. This definition keeps open the scope of the norms (i.e. who or what are the governors and the governed) as well as the range of means and ends that are appropriate to satisfying the norms (e.g. the degree of explicit coercion involved). To be sure, I have periodically offered substantive answers to these questions in the sphere of knowledge production.[2] However, the key feature of this general definition that distinguishes me from the analytic social epistemologists is that for me norms are constructed out of an already existing situation—in medias res, as it were—and *not* from a set of intuitive principles from which acceptable and unacceptable states of affairs are then deduced, typically with the aid of thought experiments. Indeed, it has been in this respect that my social epistemology has been most clearly "naturalistic," which in turn makes knowledge of history and the social sciences a precondition for any successful normative enterprise.[3]

My definition of normativity carries a presumption in favor of changing default patterns of behavior. This corresponds to how the law is normally understood—namely, as upholding a standard from which people might otherwise be inclined to deviate, were the law not present to enforce the standard. This rather "corrective" sense of normativity is also assumed by the two great theories of ethical modernism, Kant's deontology and Bentham's utilitarianism, both of which would have people improve their behavior in line with some superior conception of human rationality. (Their own conceptions differed, of course, but in exactly which respects remains an open question.)

But how does *agency* need to be understood for such a conception of normativity to work in the context of social epistemology? I explore the question, first, by identifying a form of legally recognized agency—one that bases rights and responsibilities on liability rather than property—as underwriting the change-friendly conception of normativity required by my version of social epistemology. I then zero in on the "ought implies can" principle, the Kantian slogan that serves as the benchmark in contemporary normative theory, for thinking about the sort of being we should be given our capacity for self-transformation. Finally, I turn specifically to "epistemic agency." In line with my original work in social epistemology,[4] I argue here that we need to invert the analytic epistemological understanding of the relationship of "belief" and knowledge,' treating 'belief' less as something grounded in past experience (of oneself and/or others) than as a risky projection into the future, *à la* Pascal and William James. Readers should understand this general line of argument as being in support of a broadly 'transhumanist' position, which accepts that the quest for self-transcendence is built into whatever else it means to be 'human.'[5]

1. LIABILITY AS THE SOCIAL EPISTEMOLOGICALLY RELEVANT BASIS FOR AGENCY

I take 'agent' to be a metaphysical way of talking about 'person,' a legal category that is normally but not exclusively, or even necessarily, applied to all members of *Homo sapiens*. For example, corporations are legal persons whereas the personhood of antenatal, disabled and dead members of our species remains controversial—as in the past had been various ethnic minorities, women and even children. In the past, philosophers who have been inclined to treat all humans as persons, such as Kant and Hegel, spilled much ink over the terms of entitlement to such an assignation, given the palpable empirical differences in the capacities and achievements of actual human beings. Here it is worth recalling Locke's point that 'person' is a 'forensic' category, which is to say, assigned in a context where one wishes to attribute responsibility for action. In the law, this means that a person is the bearer of 'rights' (and corresponding 'duties'), which together define the field of possible interaction between persons.

What we call a 'liberal' society is one where rights are prescribed such that everyone is free to the same extent to act. While nowadays philosophers focus on what 'free to the same extent' means, the idea was originally striking simply in its stress on default equality, the removal of any social markers from being 'free to act.' Thus, the legal person is an 'individual' mainly in the sense of being a distinct locus of rights and responsibilities from that other people who might share much in common, including a line of descent. Traditionally, with some clear and context-specific exceptions (e.g. military and business expeditions), legal systems had tied sphere of freedom to hereditary social standing. Even republican societies, which ideologically opposed the prerogatives of birth, typically demanded some proof of capacity for self-rule (e.g. property ownership) as grounds for admission as a legally recognized equal, or 'citizen'—and this often turned on some inherited advantage. For this reason, prior to the establishment of the United States in the late eighteenth century, such societies tended to remain small. In the twentieth century, the welfare state tried to implement the requisite sense of egalitarian liberty as 'equality of opportunity,' which licensed the state (via the tax system) to intervene proactively in people's lives to ensure levels of health, education, and social security sufficient for them to fully exercise their rights.

However, these moves did not address the deeper problem with the idea of a liberal society, which pertains not to the unequal starting points of persons but how to deal with the unequal consequences of their actions. Liberalism is biased towards productivity. *Prima facie* the people who are likely to do better in a liberal society are the ones who do more, which is to say, make the most of their liberty. In this respect, libertarianism is liberalism seen from the standpoint of its winners. But it is not clear that the most benefits are accrued to society—which after all was why liberalism was first proposed as an antidote to feudalism—if such a *laissez faire* policy is adopted. Many opportunities for growth may be lost, if a field's pioneers manage to capitalize on their original actions to such an extent as to crowd out future agents from entering it. This is why the antimonopoly legislation initiated in the United States during Theodore Roosevelt's presidency has been seen as a foundational moment in the distinctly American usage of 'liberalism,' which in the spirit of Hegel assumes that a strong state is the guarantor, not the opponent, of freedom in all its legally recognized forms.

It is worth noting here that the arguments for constraining agents who exert monopoly control over a field, while couched in terms of a kind of intergenerational justice between putatively equal agents, is ultimately susceptible to empirical scrutiny in terms of actual consequences. Again, if we are concerned with maximizing the benefits accrued under liberalism, one needs to investigate whether monopolistic regimes are actually less efficient and innovative than more pluralistic ones. The cerebral Silicon Valley venture capitalist Peter Thiel has recently presented a counter-intuitive but empirically based argument for the efficiency of monopolies in terms of their ability to stabilize a competitive environment, which might otherwise generate wasteful expense as rivals try to outdo each other in doing basically the same thing.[6] Thiel asks: Would it not be better for all concerned if the rivals either joined forces with the monopolist or shift their energies to a field clearly beyond the monopolist's reach? Meanwhile the monopolist can be left to do what it does best, which is to practice 'economies of scale' with regard to production and distribution as well as horizontal integration of activities related to production (e.g. supply of raw materials and distribution of finished products). Not least, the monopolist can probably also absorb the cost of bad consequences without undergoing self-destruction, an often overlooked precondition of learning. After all, errors are never freestanding events but relative to the epistemic state of the agents who commit them. Arguably, then, in order for the full corrective benefits of learning to be reaped, thereby preventing 'corporate amnesia' or loss of institutional memory (Douglas 1987), the erroneous agents should be allowed to carry on, albeit in a state constrained by knowledge of their error. Taken together, the foregoing advantages of monopolies constitute, in the economists' jargon, strategies for minimizing 'transaction costs.'[7]

Although Thiel himself might not welcome the example, the global academic system organized around a cartel of universities has historically exhibited the virtues of just such consolidated control—in this case, in the field of systematic knowledge, ranging horizontally across research, teaching, and 'outreach' functions. Thus, academic knowledge production exhibits many of the same historical patterns of dominant business firms, especially in terms of incorporating preferred raw material suppliers as a division of its operations, an 'internalized externality.' What I mean here is the assimilation of various previously freestanding forms of data gathering, from naturalists and ethnographers to public opinion pollsters and market researchers, as part of academia's normal knowledge production portfolio. However, the typical means by which this has happened—the issuing of credentials to preferred suppliers (typically involving prior displays of their methodological faith)—has in turn begotten classic liberal arguments against academia as an inherently 'rent-seeking' institution that places high entry costs (aka credentials), while trying to exact heavy tolls (comparable to papal anathema) if knowledge has not undergone its 'peer-reviewed' processes.[8] To his credit, Thiel has recently taken his own advice to heart by funding a scheme designed to take students out of the academic cartel altogether in the name of an alternative knowledge-producing field that is defined by entrepreneurship, start-up companies, and venture capital.[9]

In modern law, rights and responsibility have been normally assigned in terms of powers intrinsically possessed by the agents, as opposed to powers that are made available to the agents by virtue of a situation that mutually implicates them in some exchange. After Calabresi and Melamed's 'Property Rules, Liability Rules, and Inalienability,' a cornerstone in the law and economics literature, this is known as the distinction between *property* and *liability* as the basis for 'entitlement,' which is to say, the conditions under which an agent is licensed to act in a field inhabited by multiple agents.[10] The distinction is conceptually subtle but its practical consequences are quite profound.[11] The sea change recommended by the liability approach is, in effect, to say that even if all agents are free to the same extent, justice demands careful consideration of what combination of actions will result in both sides getting something of what they want while maintaining their capacity to remain free agents. Ironically, as we shall see, this requires that each party to an exchange thinks of itself as a potential monopolist who is at least partly responsible for the welfare of the other party with whom it engages in exchange.

For example, if we always understood copyright as *literally* a form of 'intellectual property,' then whenever a radio station wanted to play a song, it should ask for the songwriter's permission because to do otherwise would be tantamount to trespassing. Our default intuitions against 'invasion of privacy' and 'manipulation' also trade on this understanding, which imagines agents to be bounded like countries that are entitled to exert strict border controls over who is allowed entry. However, if one considers the potential volume of radio airplay, the potential benefits that it might have for the songwriter, but also more generally the spirit in which 'freedom' is legally recognized, the idea that permission must always be sought before an agent can intervene into any other agent's sphere of action seems unduly restrictive in principle. In practice, it would rack up large transaction costs, as radio stations would be forced to secure separate permission for each play of a song. However, in the case of airplay, radio stations pay a block fee upfront in anticipatory compensation for any loss revenues to the artist who no longer needs to be heard live. But the size of this fee is mitigated by a variety of indirect benefits (e.g. additional gigs, a renewed record deal) that the songwriter receives from the publicity.[12]

Such is the logic behind the liability model of agency. It should appeal to a social epistemologist—especially of a constructivist bent—because it makes one's own *de facto* agency depend on possibilities that are opened or closed by the presence of other agents. What exposes me to potential harm also exposes me to potential benefit, and both need to be taken into account when determining the terms of a just exchange—that is, '*ex ante* compensation.'[13] The question boils down to who pays what to whom in the exchange.

Calabresi and Malamed were partly inspired by—and have been used effectively in—the interpretation of environmental protection legislation in the period which brought 'risk assessment' into the policymaker's lexicon. The paradigm case is an industrial firm that has legally purchased land but its plans for the land are reasonably seen as threatening the livelihood of its neighbors. A judge would thus need to rule on the terms under which both agents are permitted to act within a space defined by mutual agreement under contract law, effectively creating an autonomous micro-regime. In the case of 'pollution rights,' the liability approach of Calabresi and Malamed countenances two possibilities: either the would-be polluter would pay its neighbors upfront for anticipated harms caused to the environment or the neighbors would pay the would-be polluter upfront for the anticipated cost of the inconvenience of having to shift sites.[14] Which version of the liability rule prevails depends on a judgement about overall welfare considerations. Moreover, the two versions of the rule are themselves open to interpretation.

In the first case, the 'payment' to neighbors need not be restricted to a bribe-like financial settlement to allow the factory to be built; it may also involve mandating ecologically sound manufacturing principles in the design and operation of the factories so as to mitigate environmental damage. The firm may be out of pocket to the same extent in the two scenarios, but the judge would decide which scenario is likely to result in greater social benefit.

In the second judicial possibility, the neighbors could bear the burden of compensation through a financial settlement. Thus, the firm would be paid to refrain from an action that is, strictly speaking, legally permitted—namely, building a factory on land it owns. But another option is for the neighbors to do what they can at their end to accommodate industrial expansion into their space. This might not require any direct financial transactions to the firm but instead involve, say, the neighbors investing in systems designed to sequester or recycle pollutants that reach their land, while the firm is issued a temporary restraining order to not start building until the neighbors have installed the various antipollution measures. This last scenario is especially interesting—and controversial—because it effectively concedes that industrial expansion should be expected and, consequently, residents in neighboring lands should be prepared to make the relevant adjustments to their holdings. (In all these scenarios, it is not unreasonable to assume that some public subsidy will be made available to bring about whatever is seen as the desirable outcome.)

While the adjudication of 'pollution rights' provides an ideal case for enforcing the liability model of agency, it is not the only one. Consider the case of labor replacement via automated technology. A firm wanting to import robots into the workplace is like the would-be polluter, while the unionized workers resisting it are like the neighbors threatened with pollution. The pretext for such a development is usually the need to increase efficiency, with workers at least nominally allowed to remain employed if they conform to the new efficiency standards. However, in practice one might expect that the workers will be pressured to meet standards to suit machine performance—and over time the humans will lose their jobs, especially if labor unions continue to lose their bargaining power. Setting aside the financial settlements that one party might make to the other, we can imagine a resolution happening in one of two general ways: On the one hand, the firm may be required to pay an 'automation tax' that goes toward retraining the workers who end up losing their jobs. On the other hand, the workers may be required to purchase insurance to cushion the blow in the event of job loss. (Again, both options may be publicly subsidized to some degree.) The example of the automated workplace shows, perhaps even more vividly than the pollution rights one, that the liability model of agency is forward-looking, presuming considerable flexibility on the part of the agents to adapt their interests to their circumstances, instead of claiming rights simply on the basis of their historic property arrangements.

Now let us imagine the liability model of agency as something that becomes ingrained into a society's sense of justice over time. At the start, one can imagine a focus on financial transactions between the relevant parties, such that the more inconvenienced party is compensated by the less inconvenienced one, resulting in each being allowed to do a version of what it wants to do. However, this still retains the property model's strong sense of territorial boundaries between the parties. Thus, the next and deeper stage in the acceptance of the liability model involves removing just this vestige of the property model by effectively requiring that each party envisage the other as a potential extension of itself—that is, in economists' terms, an 'externality' that is internalized into the agent's utility calculus. In real-world cases, as in the adjudication of royalties for airplays of a song, the settlement may make the songwriter (or whoever holds the song's copyright) and the radio station effective co-owners of— or jointly responsible for—the song in question. As a result of the settlement, both have a vested interest in promoting the song, even though the song functions differently in their respective utility calculations. In this respect, a contract may be seen as a struggle for mutual incorporation, resulting in overlapping spheres of agency.

2. THE LIABILITY MODEL AS AGENCY ENHANCING: REVISITING 'OUGHT IMPLIES CAN'

Kant famously presented 'ought implies can' as a logical presupposition of morality. He understood the slogan as expressing a categorical freedom, a permanent possibility of free will. This is in contrast to the sort of freedom that Kant attributed to Hume, so-called 'hypothetical freedom,' as in 'I could have done X, had I not chosen to do Y.' The spirit of what Kant was suggesting would seem to go very much against what economists call 'path dependency,' namely, the idea that a decision I take now may subsequently constrain the decisions that are available to me. From this standpoint, all I ever have is hypothetical freedom. For this reason economists tend to want 'opportunity costs' (i.e. what one might have done otherwise) included in a full accounting for costs: An agent's success or failure is partly determined by the range of actions from which it had to decide. As Popper famously argued, only on that basis—the 'logic of the situation'—can the agent's rationality be judged.[15] Thus, while I may have done better had a different range of options been available, in practice I did as well as I could have, given the options actually before me.

But Kant's idea of categorical freedom demands more than 'rational under a given set of circumstances': It demands 'rational under any and all circumstances.' Now, how might such a state be achieved—or even sensibly aspired to? The most obvious strategy would involve reversing any constraints that might create path dependencies in an agent's decision-making process. Assuming that we lack the time-travel capacity required to literally undo what has already been done, we might strive to achieve a state-of-affairs that the agent would regard as 'functionally equivalent' to the one of categorical freedom that one requires. As Arthur Ripstein has observed, the law increasingly turns to financial compensation to simulate just such a restoration of freedom in cases where 'path dependency' is the result of harm caused by another party.[16] The various forms of *ex ante* compensation associated with the implementation of the liability of model of entitlement may be understood in this light. In that case, judicial flexibility in the face of conflicting rights operationalizes the law's recognition of the value of categorical freedom to agents.

Nevertheless, before restoring Kant's sense of categorical freedom, we need to deal with two of the three major interpretations of Kant's 'ought implies can,' which would bring categorical closer to hypothetical freedom:

1. That you should only commit to do that which you are capable of doing.
2. You should only be held accountable for that which you are capable of having done.

Taken together, (1) and (2) restricts the scope of ethics to what virtue theorists would recognize as the 'natural' dimensions of the human condition, whether the conception of 'human nature' draws from Aristotle, Aquinas or Darwin. On this view, morality is for making us better beings of the sort we already are, not for turning us into something we are not— namely, saints or supermen. Such a position can be seen as following from a sense of hypothetical freedom that might be possessed by a believer in what can only be called 'special creationism': That is, once you have been created as member of a particular species, you are bound by the possibilities of that species and hence incapable of transcending them and, therefore, you should not be expected to do otherwise—even to meet some ideal posited by members of your species (e.g. eugenic fitness). Philosophers who would not be normally classed as 'Neo-Aristotelian,' such as Jürgen Habermas, have nevertheless tied their views of human categorical freedom to just this understanding of the human condition.[17]

I deliberately put Darwin in this camp of crypto-creationists because there is an unholy alliance between virtue theory and 'evolutionary ethics' when it comes to downsizing human moral demands to fit the default capacities of natural born members of *Homo sapiens*. In this context, our default incapacities (i.e. our biases and cognitive limitations, the diminished survival skills of the very young and very old, etc.) are invoked as a pretext for explaining the emergence and maintenance of social bonds.[18] There is little recognition in this literature that science and technology might actually *enhance* native human capacities to render us capable of meeting the high ethical demands of Kant—or, for that matter, that other standard bearer of ethical modernism, Jeremy Bentham. To be sure, Kant was originally talking about a personal ethic, while Bentham was addressing the ethic of the legislator. Yet, each in his own way attributed to humanity a 'devolved' conception of divine agency: in the one case devolution to the individual, in the other to the state. But in both cases, the aspiration remained universal normative coverage, as in God's imposition of natural law.[19] This lingering sense of divine ancestry appears in both the Kantian need for superhuman feats of will and foresight to grant everyone else the same license we grant ourselves and the Benthamite need for superhuman feats of calculation and forbearance to enable people to engage in long-term welfare maximization.

As Hannah Arendt perceptively observed, once Kant's political philosophy is taken into account, his overall normative position is, perhaps surprisingly, not so far from Bentham's.[20] After all, while Kant admitted that each human possesses a dignity that entitles him to be treated as an end in himself, he also believed in human progress. And from this standpoint, each person is simply a moment in the perfection of the species. This tension between humans regarded as ends and as means was subsequently exploited by Hegel and Marx under the rubric of the 'cunning of reason.' Kant had appeared to resolve the matter to his own satisfaction by abstracting the concept of humanity as a regulative ideal from the various imperfections of human societies—indeed, of the sort that he detailed in his controversial late work, *Anthropology*. It is only under the guise of 'humanity' in this ideal sense—as members of the 'Kingdom of Ends'—that Kant accords us dignity. In contrast, he treated actual embodied and embedded humans as occupying positions in a hierarchy and waystations on the collective journey to the Kingdom of Ends.

For his part, although Bentham disavowed any transcendental sense of humanity, he joined Kant in rejecting tradition and common sense as providing a reliable normative compass. In effect, both recognized a productive tension between 'is' and 'ought' (or 'matter' and 'spirit'), where 'can' performs a mediating role. To be sure, their metaphysical starting points were quite different. Kant thought about humanity in *many-one* terms, Bentham in *part-whole* terms. Thus, whereas Kant was inclined to see individual humans as better or worse representations of the ideal of humanity, Bentham regarded them as coproducers of general welfare, albeit not necessarily in the proportions that they thought. (Here it is worth recalling that 'utility' was originally presented as an *objective* measure—modelled on money—of what people experienced as pleasure and pain, a view that was overturned by the subjective 'marginalist' turn in late ninteenth century economics.) But both positions remain indebted to the divine legacy, insofar as their normative perspective requires a second-order understanding of agents from outside the field of action.

Here it is worth observing that the distinction between means and ends is never so clear to agents operating inside a field of action. Thus, members of a given society tend to identify with their most enduring practices, as if how a society arrived at its present state were a reliable guide to a future that is bound to present new challenges. To rise above this perspective, one needs to legislate relatively unencumbered by the past, as per Kant's categorical freedom or Bentham's radical reconstruction of parliamentary practice. This point goes to the difference in the role that *science* plays in Kant and Bentham vis-à-vis virtue theorists and evolutionary ethicists. The latter pair are more informed by the empirical findings of science, especially from natural history, whereas Kant and Bentham took their lead from the scientific method itself, especially the way that reason and experiment have revealed new possibilities for action, which might conform to a superior ethical standard. In short, Kant's deontology and Bentham's utilitarianism jointly oppose the idea that one's normative horizons are simply acquired from others and then passed on as default patterns of social life. On the contrary, they are self-consciously constructivist and even revisionist with regard to morality. This joint refusal to let the future of humanity be dictated by its past most clearly marked Kant and Bentham as philosophers who, unlike the virtue theorists and evolutionary ethicists, do not let 'ought implies can' to slide into 'is implies ought,' as in (1) and (2) above.

Kant and Bentham came to be presented as opponents in the mid-ninteenth century imagination (e.g. John Stuart Mill, Henry Sidgwick) only once their normative principles were seen as applying to individuals living in a democracy, that is, a society where anyone is a potential legislator, thereby blurring the boundary of morals (Kant) and politics (Bentham). Nevertheless, the need for god-like powers persisted as the legal system increasingly recognized the individual *as such* as the ultimate locus of rights and responsibilities, and hence of agency in society. At that point, the individual became the paradigm case of the legal person. Subsequently, such collectively and inter-generationally defined 'artificial persons' as states and firms came to be seen as deriving their social legitimacy—albeit not their formal legal powers—from the decisions taken by individual persons.

If the high ethical demands of Kant and Bentham are worth meeting, then why should we not enhance ourselves accordingly—even if that forces us to radically rethink our social bonds, such that *autonomy* rather than *dependency* becomes the preferred basis for social relations? After all, as Kant made explicit in his famous essay that popularized the term 'Enlightenment' in German (*Aufklärung*), the maturation of humanity as a species rests on the gradual removal of those features of social life that keep us in a childlike state longer than is biologically necessary. In the late eighteenth century, this was read as an attack on 'paternalism' wherever it was practiced—from the home to the state. In our own time, it would probably also encompass 'expertise,' insofar as it compels deference to those with greater experience in a given domain, thereby inhibiting the capacity to judge for oneself. I shall return to this point in the next section, specifically on 'epistemic agency.'

But more to the point here, modern accounts of 'empowerment'—of the sort that have fueled revolutionary movements since the English Civil War in the seventeenth century—have typically involved people realizing that they have hidden (originally, God-given) powers that are being arrested by existing social arrangements. In this respect, the sorts of 'checks and balances' prescribed in the U.S. Constitution, perhaps the Enlightenment's most enduring concrete legacy, can be understood as using the legal system to specify a nation-wide 'smart environment' that enabled individuals to realize their potential, presumably for the greater welfare of all—or 'commonwealth,' in the preferred term of the time. Our own times face a somewhat different challenge to empowerment, which might be called 'second-order paternalism,' involving prohibitions on, say, tinkering with the human genome or radical experimentation on human subjects. Here we need a new constitution relating to 'scientific citizenship' that insures the liberties needed on all sides to pursue these adventurous new developments.[21]

To be sure, in both the historical and the contemporary cases, there is a certain 'blind faith' in the ultimately beneficial character of the changes needed to unleash hidden potential, which in the historical cases led to significant distress and harm and often fell short of aspirations. Yet, it is precisely here that the liability basis for ascribing rights and responsibility becomes relevant: Some combination of *taxation* and *insurance* can mitigate transitional harms caused to the parties in question. But both of the terms highlighted in the previous sentence need to be understood in a fairly imaginative fashion. For example, taxes paid by an innovator capable of causing harm (e.g. as a 'polluter') may be hypothecated to concerns directly relevant to those likely to be in harm's way in this and similar cases. Correspondingly, those in harm's way may take out insurance in the form of 'upgrading' themselves or their possessions so as to become more resilient to harm, should it come their way, either now or in the future.

At this point, we arrive at what I believe is the correct interpretation of Kant's sense of categorical freedom, which compels us to be the sort of beings who can live up to the correct moral principles, as in the following 'proactionary' reading of 'ought implies can':

1. That if you want someone to do something that they cannot do, then you must provide the means by which they are made capable of doing it.

This properly imperative way of reading Kant's slogan leaves open who exactly is to do the relevant 'providing'. But one thing is clear: The way we are is not necessarily the way we need to be in order to do what we ought to do.

3. WHAT IS SO 'EPISTEMIC' ABOUT EPISTEMIC AGENCY?

The concept of epistemic agency is founded on an ambiguity: Is it about the epistemic character of all forms of agency, or about a specific form of agency that pertains to epistemic activities? Or perhaps both (or even something in between), if one believes that all forms of agency are intrinsically epistemic. The third option is the one that makes the most sense to me for reasons that will become apparent from the following example.

Nelson Goodman's student Catherine Elgin has imported 'epistemic agency' into the education literature as a species of virtue theory, where it is tied to the idea of taking responsibility for one's beliefs—a sense of agency associated with specifically epistemic activities.[22] In developing this understanding of the concept, she stresses the need to learn respect for expertise, which is in turn tied to the self-discipline required to not simply believe what one wants to believe. While it is easy—and perhaps even true to Elgin's intentions—to see her as trying to stem a perceived wave of subjectivism and relativism among contemporary students, what is most philosophically interesting about her analysis is its deployment of Kant to make the point.

Elgin assumes that Kant's Enlightenment imperative to think for oneself—'owning' one's beliefs—entails (1) understanding how the experts have reached their conclusions, and *as a result* (2) believing those conclusions, unless one has done relevant independent research. While (1) seems quite reasonable preparation to taking a stand on an issue, (2) certainly does not follow from it. In effect, Elgin's conception of epistemic agency entails that one engage in a kind of sympathetic understanding (*Verstehen*?) with experts that her argument would seem to deny to, say, members of particular interest groups, native cultures, etc., with whom one might otherwise spontaneously identify. Surely, for epistemic agency to remain properly 'agential,' it must allow for the prospect of people becoming competent in a particular point of view yet in the end rejecting it as false. To be sure, this then opens a Pandora's box of issues of how to deal with the consequences of many competent people equipped with the same beliefs drawing radically different conclusions.

In any case, Elgin should not find the result surprising, since it had been already adumbrated by her Harvard mentors Goodman and Quine as (in the latter's terms) the 'underdetermination of theory choice by data.' However, they adopted a conservative presumption that we should err on the side of expecting that the future will be like the past (cf. Fuller and Collier 2004: chap. 10). Consider Goodman's famous thought experiment,[23] which would have us consider the possibility of two properties 'green' and 'grue,' which apply equally well to all emeralds before a given time but after that time, 'green' proponents say the emeralds remain green, while 'grue' ones say they turn blue—not because the composition of the emeralds themselves change but because we discover that they had been 'grue' all along.[24] Nevertheless, the rhetorical force of Goodman's thought experiment is that 'grue' remains a far-fetched option even though the concept captures everything that we currently know about the color of emeralds.

In Elgin, this conservative presumption is manifested as an abiding respect for expertise, the knowledge possessed by the established authorities. But a Goodman-inspired counterpoint applies here too: Even granting a consensus among experts now, it is easy to project a future that violates this consensus, not simply because the future presents circumstances unforeseen by the experts but because (*à la* Goodman) these unforeseen circumstances provide new opportunities to revisit the truth of what the experts had been saying all along. Nevertheless, as we have also seen, Elgin's notion of 'belief' is much too loaded, combining the entertainment of a proposition and the disposition to act on it. As a result, one is left with the misleading impression that holding a belief somehow compels one (rationally?) to act a certain way. It is as if to build the coercive character of expert authority into what *prima facie* is simply an expression of the cognitive content of expert knowledge.

The problem is very much in evidence in the 'evolution' of creationism to intelligent design theory as an alternative to the Neo-Darwinian synthesis in biology. As time goes on, it becomes increasingly difficult to dismiss the opponents to Darwin out of hand, as they become more competent in the relevant established sciences yet still insist on spinning them in their own favor. In this context, the analytic philosophical tendency to speak generically of epistemic states as 'beliefs' is profoundly misleading. 'Information' would be a more accurate term in these cases. Thus, when students learn Neo-Darwinian biology, they are acquiring in the first instance not *beliefs* but *information*, which they may then either convert into 'beliefs' in the sense of making them the basis of their own assertions—or not, as the case may be. In the latter case, they keep the information in reserve for redeployment in the future, perhaps in order to deconstruct the knowledge base from whence the information came. But often something subtler is going on. A good, albeit controversial, case in point is Stephen Meyer's *Signature in the Cell*, which takes much more literally than molecular biologists would perhaps like (or expect) that the code-like character of the 'genetic code' implies the existence of a codemaker (aka The Intelligent Designer aka God).[25]

I have recently written about this general development as the 'customization' of science, drawing on a distinction in the marketing literature between *customer* and *consumer*.[26] The former, often a retailer, purchases goods ('wholesale') not for personal use but for resale. The 'reselling' adds to the value chain, as the retailer is forced to customize the product to prospective consumers. Minimally this requires that the retailer is a more efficient distributor of goods than the wholesale supplier. However, retailers have increasingly added value by heightening and/or obscuring certain features of the goods that expand their markets beyond what wholesalers had envisaged. Think about it this way: A fancy restaurant and a fast food chain may purchase their meat from the same wholesale agent who delivers 'Grade A' beef, but what the consumer ends up getting is significantly different in the two cases—but not necessarily in the sense that one is 'better' or 'truer' than the other.

In this respect, intelligent design theorists may be seen as enterprising retailers who have found new uses for much of conventional biology, which in many cases goes against the spirit of the epistemic wholesalers (aka experts). Of course, wholesalers have been known to stop trade with various retailers on moral and political grounds (and vice versa). In a similar spirit, the likes of Richard Dawkins and Daniel Dennett favor decontaminating students of religious beliefs before entering a science class.[27] However, as long as the values of liberal democracy trump those of expert authority, the customization of knowledge is unlikely to disappear soon—in fact, it is clearly on the rise.[28]

My original formulation of 'social epistemology' offered a somewhat paradoxical way of thinking about the matter, namely, that what we acquire in the biology classroom, say, is *knowledge* but it is up to us to decide whether to *believe* it.[29] Implied here was that knowledge is justified through a kind of organized social construction, modelled on the practices of publicly accountable academic disciplines. But this 'knowledge' is ultimately no more than a common inheritance or capital, which may be invested in many different ways, with those investments properly constituting 'beliefs.' In that case the interesting question to ask is whether the returns on those investments result in the production of new knowledge. This in turn has led me to embrace a vision of epistemology as 'cognitive economics.'[30]

Given this understanding, epistemologists can be seen to have classically focused on *licensing* rather than *insuring* beliefs, which in turn should not be confused with *protecting* beliefs. A 'protected' belief is one guaranteed some measure of epistemic standing regardless of its encounters with reality because of the status of the believers. Multiculturalism is perhaps the most natural context nowadays to envisage this situation, whereby a near-identity is presumed between beliefs and their believers.[31] However, in the 1970s—largely under the influence of Karl Popper's student Imre Lakatos—there was much discussion of science as itself an entity that devoted some of its epistemic resources to self-protection via what Lakatos called the 'negative heuristic,'[32] which (rightly in Lakatos's view) licensed the continuation of a research programme's 'metaphysical hard core' despite having generated anomalous empirical results. In contrast, in the case of an 'insured' belief, an encounter with reality may result in a significant change of beliefs, yet without irreversible damage done to the believers. This is truer to Popper's original falsificationist ethic, which was about encouraging scientists to take more, not fewer, risks at a time when science was becoming increasingly dominated by consensus-based thinking of the sort championed in Thomas Kuhn's *The Structure of Scientific Revolutions*.[33] In that case, shedding beliefs is more like the normal organic process of shedding skin than a potential threat to the life of the organism. Indeed, Popper generally avoided the word 'belief' in discussions of scientific epistemology, preferring instead 'hypothesis,' which stresses the disposable character of the scientist's epistemic assertions.

Against this backdrop, it is interesting to return to the Biblical source of what the sociologist Robert Merton dubbed the 'Matthew Effect,' which alludes to Jesus' parable of the talents in Matthew 25.[34] ('Talent' originally referred to units of wealth but the term's remit expanded over time to include the capacity needed to possess such wealth.) Merton appealed to the parable to articulate the principle of cumulative advantage in knowledge production, a positively spun version of path dependency. For our purposes, it provides institutional expression to the Quine-Goodman-Kuhn line that a conservative normative presumption favors future epistemic practice reinforcing past practice, until proven otherwise. 'Proving otherwise' was the role that Kuhn assigned to a 'crisis' within the dominant paradigm that then licenses the sort of philosophical soul-searching eventuating in a scientific revolution.[35] However, in the original parable, the one of three sons who achieves the most with the talents he inherits is the one who invests—rather than hoards—them. In other words, the son who invested the most also reaped the most benefits, while Jesus castigates the son who hoarded his talents out of fear of losing everything. Whereas Merton stressed the size of return on investment, the parable itself seems to emphasize the role of risk taking in opening up opportunities. Popper would be pleased with this more faithful reading.

As it turns out, the idea that beliefs are risky investments that may require some form of insurance goes against a line of analytic philosophical argument, associated with Bernard Williams, which dismisses on conceptual grounds the very idea that beliefs require decisions, since beliefs are by definition 'compelled' by what one takes to be evidence—a formulation the keeps open whether the evidence is normatively as well as psychologically compelling.[36] Interestingly, Williams's argument was launched against the 'will to believe' thesis of William James, who in order to overcome chronic depression applied a prototype for 'positive thinking' he called 'meliorism,' which involves seeing the world as containing the potential to be better than it currently is. When I first confronted James's thesis, I treated it as an updated secular version of Pascal's Wager, which really involves a decision to change the environment in which one's beliefs are formed—so as to be open to evidence for God's existence, were it to present itself—and not an outright decision to believe in God.[37] This of course is compatible with Williams's argument against 'deciding to believe.' However, I have come to believe that this distinction is more semantic than real, especially once beliefs are seen as akin to investments whose value matures over time—ideally to be generative of knowledge—rather than exclusively owned pieces of mental property.

NOTES

1. Steve Fuller, *The Knowledge Book: Key Concepts in Philosophy, Science and Culture*, (Durham, UK: Acumen, 2007), 110–14; cf. Stephen Turner, *Explaining the Normative*, (Cambridge, UK: Polity, 2010).

2. E.g. Steve Fuller, *The Governance of Science* (Milton Keynes, UK: Open University Press, 2000).

3. Steve Fuller, "Epistemology Radically Naturalized: Recovering the Normative, the Experimental, and the Social." In *Cognitive Models of Science, Minnesota Studies in the Philosophy of Science*, vol. 15, ed. Ronald N. Giere (Minneapolis: University of Minnesota Press, 1992), 427–59.

4. Steve Fuller, *Social Epistemology* (Bloomington, IN: Indiana University Press, 1988).

5. 'Transhumanism,' a term coined by evolutionary biologist Julian Huxley in the 1950s, refers to a range of movements dedicated to human enhancement, a project typically characterized in terms of turning evolutionary processes to our own species' advantage. The background philosophy and politics that inform this movement is presented in Steve Fuller and Veronika Lipinska, *The Proactionary Imperative: A Foundation for Transhumanism* (London: Palgrave Macmillan, 2014).

6. Peter Thiel, *Zero to One* (New York: Random House, 2014).

7. Ronald Coase, "The Nature of the Firm," *Economica*, 4, 16 (1937): 386–405.

8. Steve Fuller, *Knowledge Management Foundations* (Woburn, MA: Butterworth-Heinemann. 2002): chap. 1.

9. Beth McMurtrie, "The Rich Man's Dropout Club." *Chronicle of Higher Education.* 8 February (2015).

10. Guido Calabresi and A. Douglas Malamed, "Property Rules, Liability Rules, and Inalienability." *Harvard Law Review*, 85 (1972): 1089 ff.
11. Fuller and Lipinska, *Proactionary Imperative*, chap. 4.
12. Lawrence Lessig, *The Future of Ideas* (New York: Random House, 2001): chap. 7.
13. Richard Posner, Economics and Justice (Cambridge, MA: Harvard University Press, 1981): chap. 4.
14. Calabresi and Malamed, 'Property Rules.'
15. Karl Popper, *The Poverty of Historicism* (London: Routledge & Kegan Paul, 1957).
16. Arthur Ripstein, "As If It Never Happened." *William & Mary Law Review*, 48, 5, (2007): 1957–1997.
17. Jürgen Habermas, *The Future of Human Nature* (Cambridge, UK: Polity, 2002).
18. Alasdair MacIntyre, *Dependent Rational Animals* (South Bend, IN: University of Notre Dame Press, 1999); cf. Peter Singer, *A Darwinian Left* (London: Nicolson and Weidenfeld, 1999).
19. Jerome B. Schneewind, "The Divine Corporation and the History of Ethics." In *Philosophy in History: Essays in the Historiography of Philosophy*. eds. Richard Rorty, Jerome. Schneewind and Quentin Skinner (Cambridge, UK: Cambridge University Press, 1984) 173–92.
20. Hannah Arendt, *Lectures on Kant's Political Philosophy* (Chicago, IL: University of Chicago Press, 1982).
21. Fuller and Lipinska, *The Proactionary Imperative*.
22. Catherine Elgin, "Epistemic Agency." *Theory and Research in Education*, 11, 2, (2013): 135–52.
23. Nelson Goodman, *Fact, Fiction and Forecast* (Cambridge, MA: Harvard University Press,1955).
24. Steve Fuller, *Preparing for Life in Humanity 2.0* (London: Palgrave Macmillan, 2012): chap. 1.
25. Stephen Meyer, *Signature in the Cell* (New York: HarperCollins, 2009).
26. Steve Fuller et al., *The Customization of Science* (London: Palgrave, 2014).
27. E.g., Daniel Dennett, *Breaking the Spell: Religion as a Natural Phenomenon* (London: Allen Lane, 2006).
28. Fuller et al. *Customization of Science*.
29. Steve Fuller, *Social Epistemology* (Bloomington, IN: Indiana University Press, 1988).
30. Steve Fuller, *Knowledge: The Philosophical Quest in History* (London: Routledge. 2015): chap. 1.
31. Steve Fuller, *The Governance of Science* (Milton Keynes, UK: Open University Press, 2000): chap. 4.
32. Imre Lakatos, *Methodology of Scientific Research Programmes* (Cambridge, UK: Cambridge University Press, 1976).
33. Thomas Kuhn, *The Structure of Scientific Revolutions* (Chicago, IL: Chicago University Press, 1970).
34. Robert Merton, *The Sociology of Science* (Chicago, IL: University of Chicago Press, 1973).
35. Thomas Kuhn, *Structure*.
36. Bernard Williams, "Deciding to Believe." in *Problems of the Self* (Cambridge, UK: Cambridge University Press, 1973): chap. 9, 136–51.
37. Steve Fuller, *"Bounded Rationality in Law and Science"* (Ph.D. diss., University of Pittsburgh, 1985): chap. 2.

REFERENCES

Arendt, Hannah. *Lectures on Kant's Political Philosophy*. Chicago: University of Chicago Press, 1982.

Calabresi, Guido. and Melamed, Douglas. "Property Rules, Liability Rules, and Inalienability." *Harvard Law Review* 85 (1972): 1089.

Coase, Ronald. "The Nature of the Firm." *Economica* 4, 16 (1937): 386–405

Dennett, Daniel. *Breaking the Spell: Religion as a Natural Phenomenon.* London: Allen Lane, 2006.

Douglas, Mary. *How Institutions Think.* Syracuse, NY: Syracuse University Press, 1987.

Elgin, Catherine. "Epistemic Agency." *Theory and Research in Education* 11, 2 (2013): 135–52.

Fuller, Steve. *Bounded Rationality in Law and Science* (Ph.D. dissertation). Department of History & Philosophy of Science: University of Pittsburgh, 1985.

Fuller, Steve. *Social Epistemology.* Bloomington, IN: Indiana University Press, 1988.

Fuller, Steve. "Epistemology Radically Naturalized: Recovering the Normative, the Experimental, and the Social." in *Cognitive Models of Science, Minnesota Studies in the Philosophy of Science*, vol. 15, edited by Ronald Giere, 427–59. Minneapolis: University of Minnesota Press, 1992.

Fuller, Steve. *The Governance of Science.* Milton Keynes. UK: Open University Press, 2000.

Fuller, Steve. *Knowledge Management Foundations.* Woburn, MA: Butterworth-Heinemann, 2002.

Fuller, Steve. *The Knowledge Book: Key Concepts in Philosophy, Science and Culture.* Durham, UK: Acumen, 2007.

Fuller, Steve. *Preparing for Life in Humanity 2.0.* London: Palgrave Macmillan, 2012.

Fuller, Steve. "Customised Science as a Reflection of Protscience." In *The Customization of Science,* edited by Steve Fuller, Mikael Stenmark, and Ulf Zackariasson. 158–75. London: Palgrave, 2014.

Fuller Steve. *Knowledge: The Philosophical Quest in History.* London: Routledge, 2015.

Fuller, Steve. and Collier, Jim. *Philosophy, Rhetoric and the End of Knowledge.* 2nd ed. (Orig. 1993, by Fuller). Mahwah, NJ: Lawrence Erlbaum Associates, (2004).

Fuller, Steve. and Lipinska, Veronika. *The Proactionary Imperative.* London: Palgrave, (2014).

Fuller, Steve.; Stenmark, Mikael.; Zackariasson, Ulf.; eds. *The Customization of Science* London: Palgrave, 2014.

Goodman, Nelson. *Fact, Fiction and Forecast.* Cambridge, MA: Harvard University Press, 1955.

Habermas, Jürgen. *The Future of Human Nature.* Cambridge, UK: Polity, 2002.

Kuhn, Thomas. *The Structure of Scientific Revolutions.* Chicago, IL: Chicago University Press, 1970)

Lakatos, Imre. *Methodology of Scientific Research Programmes.* Cambridge, UK: Cambridge University Press, 1976.

Lessig, Lawrence. *The Future of Ideas.* New York: Random House, 2001.

MacIntyre, Alasdair. *Dependent Rational Animals.* South Bend, IN: University of Notre Dame Press, 1999.

McMurtrie, Beth. "The Rich Man's Dropout Club." *Chronicle of Higher Education.* 8 February, 2015.

Merton. Robert. *The Sociology of Science.* Chicago: University of Chicago Press, 1973.

Meyer, Stephen. *Signature in the Cell.* New York: HarperCollins, 2009.

Popper, Karl. *The Poverty of Historicism.* London: Routledge & Kegan Paul, 1957.

Posner, Richard. *Economics and Justice.* Cambridge, MA: Harvard University Press, 1981.

Ripstein, Arthur. "As If It Never Happened." *William & Mary Law Review* 48, 5 (2007): 1957–1997.

Schneewind, Jerome. "The Divine Corporation and the History of Ethics." In *Philosophy in History: Essays in the Historiography of Philosophy,* edited by Richard Rorty, Jerome Schneewind and Quentin Skinner, 173–92. Cambridg,e UK: Cambridge University Press, 1984.

Singer, Peter. *A Darwinian Left.* London: Nicolson and Weidenfeld, 1999.

Stern, Robert. "Does 'Ought' Imply 'Can'? And Did Kant Think It Does?" *Utilitas* 16, 1 (2004): 42–61.
Thiel, Peter. *Zero to One*. New York: Random House, 2014.
Turner, Stephen. *Explaining the Normative*. Cambridge, UK: Polity, 2010.
Williams, Bernard. "Deciding to Believe." in *Problems of the Self*, Chapter 9. (Cambridge, UK: Cambridge University Press, 1973), 136–51.

II

Responses and Further Considerations

THREE

Two Kinds of Social Epistemology and the Foundations of Epistemic Agency

Finn Collin

In this chapter, I first sketch out a framework in which to situate the contributions from Sanford Goldberg and Steve Fuller, thus tying together two texts that, at first sight, seem very different, both in topic and style of argument. Indeed, I shall argue that, when put end to end, they constitute a fairly seamless argument. I also examine the contributions made by each to the development of a notion of epistemic agency useful for social epistemology. I conclude that, in the end, they both leave important aspects of this notion undetermined. These lacunae pertain both to the precise function of "mindedness" (consciousness) in social epistemology and the problem of nonindividual epistemic agents.

First, let us briefly review the fundamental tenets of social epistemology that constitute the common ground between the two papers. As its name indicates, social epistemology adopts a *social* perspective that no longer conceives of knowledge exclusively as an individual possession, but as something equally generated by, and attributed to, such social entities as groups, businesses, public institutions, and entire societies. Secondly, social epistemology is a *normative* enterprise, concerned not only with describing our collective epistemic practices and institutions, but also with assessing their efficacy in generating knowledge. Such critical assessment is not intended to be a mere academic exercise, but is supposed to provide directions for the improvement of those practices and institutions. Thirdly, social epistemology adopts a *naturalistic stance*, in the recognition that we cannot intuit the principles of effective knowledge generation on a societal scale by *a priori* methods. Instead, we have to take an empirical approach, drawing on such disciplines as sociology, anthropology, and economics.

Beyond this consensus, however, there are within the discipline disagreements that split it into two roughly demarcated camps. Following an emergent terminological practice, I shall refer to them as analytic social epistemology (ASE) and critical social epistemology (CSE), respectively. Our two contributors represent each of those camps, Goldberg the former and Fuller the latter. I shall now briefly characterize their differences.

1. THE TWO KINDS OF SOCIAL EPISTEMOLOGY

ASE is a descendant of classic analytic epistemology, which investigates the principles of justification of beliefs. Given a proposition that is held to be true, the task is to analyze the ways in which this supposition can be put beyond doubt and granted the status of *knowledge*. This is not a case of everyday practical doubt, however, but of defeating the arguments of the *philosophical sceptic,* who challenges even our most basic sources of knowledge, such as sensory experiences. We have no way, the sceptic alleges, to show conclusively that these experiences can be trusted; perhaps they are just illusions generated by the machinations of an evil demon, or by experiments performed on our brains by mad neuroscientists. If we accept the rules of play imposed by the radical (Cartesian) sceptic, traditional epistemology puts us at a grave disadvantage with respect to gaining knowledge about the world. What we have to work with in our cognitive endeavors are just our own subjective sensations. This threatens to result in an epistemic solipsism: even the existence of the external world is moot and, *a forteriori,* so is the existence of other human beings (minds) with whom one might have shared the task of getting to know that world. Thus, classic analytic epistemology is tied to a radically individualist perspective.

CSE, on the other hand, regards philosophy of science, rather than epistemology, as its most important progenitor. This discipline deals with an especially eminent example of social cognition, science, which by definition is a collective knowledge-seeking enterprise. Thus, while the epistemic touchstone of traditional analytic epistemology is the certainty of one's subjective experiences, philosophy of science looks to a certainty warranted by intersubjective agreement among experts, by stringent standards of experimental testing, by careful peer reviewing of research publications and—in brief—by all the institutional safeguards of science. Thus, CSE belongs to an academic line of descent in whose DNA a social perspective on knowledge was inherent from the very start. This leads to an important difference in argumentative strategy between the two schools. While CSE rests confidently in its collectivist stance, much ASE work still revolves around the legitimacy of the very step from individualism to collectivism. (The lines of demarcation are not absolute, of course; several representatives of ASE share the same collectivist point of departure, especially those whose work pertains to the sciences.)

Moreover, philosophy of science typically is not much concerned with securing indubitable evidence for beliefs already held, but rather with articulating principles for selecting what (that is, which theory) to believe in the first place. Philosophy of science generally is dedicated to a dynamic goal: promoting the progress of science. Through a number of historical intermediaries, this has engendered another key difference between ASE and CSE. In CSE, the notion of "progress" in science gradually has become an object of critical scrutiny in its own right, not only from the purely logical or semantical perspectives familiar from ASE, but also from political and ideological perspectives. Indeed, this is the main import of the term *"critical social epistemology."*

Here is the historic development in brief. Logical positivists conceived of scientific progress as the steady accumulation of data, followed by cautious generalization; the metaphor was that of adding ever more bricks to a permanent, constantly rising building. In Popper and his followers, an element of discontinuity would appear; in the progress of science, the edifices of scientific knowledge occasionally would be razed to the ground. Still, a conception of the "growth of science" was preserved, defined as the ever-closer approximation to some comprehensive truth (the notion of "verisimilitude").[1]

The view of scientific development as discontinuous was radicalized by Kuhn and his doctrine of "scientific revolutions" and "paradigm shifts."[2] But even Kuhn would subscribe to the view of science as progressive, although not as approximating some final state. To him, progress is a matter of the ever-growing capacity of science to solve puzzles. Another novel feature in Kuhn's work was his sociological angle, focusing in particular on the social structure of the scientific community.

When Kuhn's revolutionary and sociological conception of science was introduced into the cultural and political climate of the 1960s and 1970s, dominated by worries about the societal impact of science and technology, it helped spawn another, even more radical approach to science that has left a profound mark on CSE. This is Science and Technology Studies (STS), which is the critical sociological investigation of science.

A familiar perspective in sociology depicts society as an arena of struggle between societal segments with opposing interests; this conflictual perspective looms large in STS analyses of the generation and distribution of knowledge. Knowledge is power and, since power is always contested, we also may see knowledge as an object of contestation.[3] Thus, a distinctive political and moral element is introduced into the analysis of science.

The same stance clearly is evident in STS's heirs in CSE, who were never naïve about invocations of "knowledge" or "expertise" in a societal context, but insisted that such claims must always pass a scrutiny of the social forces and interests at work behind them. It is part of the remit of social epistemology to pinpoint these forces and assess their effect upon our knowledge-producing institutions.[4]

CSE is thus inherently critical and potentially revisionary with respect to existing epistemic practices. It is concerned with improving upon those practices, or even replacing them with others, if we find them more satisfactory. This revisionary tendency often assumes a clearly ideological slant, with evident sympathies for the agendas of current socio-political movements such as feminism, anti-racism, environmentalism and others. With this plurality of concerns, CSE shows an affinity with current post-modernist thought; it marks a clear divergence from the hegemonistic, reductionist tendency of classical Marxist "critical sociology."

In ASE, the break with the Cartesian tradition went hand in hand with the emergence of an alternative, *externalist* conception of human knowledge. As externalists point out, elements independent of the epistemic subject's "mind" and conscious epistemic calculations may be relevant for the epistemic assessment of the latter's beliefs. The classic example is the sensory system. A person's conviction that there is a flower vase on the table may count as knowledge, if it is produced by her visual system under favorable conditions of observation—even though she may have no clue about the mode of operation of the visual system and no precise statistics concerning its reliability. All that is required is that the system is *actually* reliable. A key figure in these developments was Alvin Goldman.[5]

In a subsequent step, Goldman would extend externalism also to encompass the social elements of the individual cognizer's environment.[6] Like CSE, the approach adopted would be naturalistic, relying crucially on empirical investigation of epistemic practices. It would also be normative; its goal not merely description, but improvement of existing practices. Still, there would be a characteristic difference in stance. ASE tends to be implicitly conservative and cautious in its approach, carefully comparing existing cognitive practices to identify the most reliable ones, while avoiding speculations concerning more radical transformations of our epistemic practices. Likewise, ASE would take a great interest in the concept of expertise and discuss principles for selecting the most competent experts, while never questioning the very institution of granting special authority to select individuals.

2. THE MATERIAL TURN AND EPISTEMIC AGENCY

In both wings of social epistemology, the extension of epistemology to include social collectivities, as well as physiological processes, instruments, and other material props, soon would be followed by an even more radical step: the elimination of the very distinction between these kinds of epistemic factors. This is known as the "material turn," a term that is somewhat misleading, as it suggests that the epistemic workings of "minds" are reduced to material processes as standardly conceived. Rather, the ambition, especially in CSE, was to devise an entirely new vocabulary, in order to describe the roles of minded and material components of cognition in the very same conceptual categories. Characteristically, that turn first occurred within CSE—and in a more radical version. A key figure in the material turn was Bruno Latour, a leading representative of Science and Technology Studies, with his theory of "actants."[7]

Latour's work on actants is part of a large-scale revolt against what he refers to as *Modernism* in Western thought.[8] Modernism is defined *inter alia*, by upholding a strict subject-object division in epistemology and a firm mind-matter distinction in metaphysics. Latour dismisses both these Cartesian dualisms in favor of a radically instrumentalist theory of (scientific) knowledge and a constructivist, anti-realist metaphysics. Scientific theories, like technology, are ways of organizing humans and material objects into complex social practices. (Latour proposes that we refer to science and technology together under the term *technoscience*, to signal their fundamental unity.) Scientific knowledge is not a passive reflection of reality in scientists' minds, but rather a collective way of coping with reality within the framework of sophisticated socio-material practices. In place of a metaphysics (and epistemology) where subjects (minds) and objects (material things) confront each other nakedly, with no intermediaries to connect them, Latour suggests a world where only intermediaries exist. No thing exists "objectively" (that is, independently of particular materially embodied cognitive practices in which it is enveloped); conversely, there are no subjects—no pure *res cogitantes*—defined as the passive representers of such objects. What we call "subjects" are only further elements of such practices. A strictly dualist ontology is thus replaced by a monist ontology, where only one kind of (fundamental) entity exists, which combines and recombines in ever-changing networks. Latour refers to these entities as "actants."[9]

There is a similar, but, characteristically, somewhat less radical, movement towards extended cognition and materialism in ASE. A seminal impetus was delivered in an article by Andy Clark and David Chalmers, "The Extended Mind."[10] They argue that there is no warrant for distinguishing between the way we attribute knowledge about the past to a person relying on his memory or to a person relying on a material prop such as a written diary. For dramatic effect, the authors envisage an

Alzheimer's patient, who relies entirely on such a cognitive prosthetic for knowledge of his own past.

The authors argue that these two sources are on a par from an epistemological point of view (provided they are equally reliable). Indeed, they, like Latour, go on to argue that the case undermines the very distinction between minded and non-minded (roughly: non-conscious) cognitive processes: That difference is not a principled one.

3. GOLDBERG ON EPISTEMIC AGENCY

After having thus placed Goldberg and Fuller in the larger context of current social epistemology, I shall now take a closer look at their contributions to the present volume. The main objective of both authors is to work out the principles of an epistemology that embraces a social perspective and to outline directions for future work. A second theme, more implicit than explicit, is an effort of demarcation in regard to the materialist turn in social epistemology. With respect to both issues, the notion of epistemic *agency* plays a key role. Let us start by looking at Goldberg's argument.

Goldberg defines social epistemology as the systematic exploration of *the epistemic significance of other minds* (chapter 1, section 1; Goldberg's italics). According to the author, this is best brought out by conceiving of epistemic subjects as *agents* who interact with other similar agents. The explicit mention of *minds* in the definition suggests a desire to maintain a distinction between the epistemic roles of (minded) human beings and measuring apparatuses (and other pieces of machinery); this is confirmed by Goldberg's further argument. How does he justify his refusal to follow along with the materialist turn? After all, the point of epistemic externalism is precisely to make room for the epistemic role of machines and other material items, if they are reliable indicators of relevant facts. Why is the possession of a *mind* a necessary condition of epistemic agency?

For pedagogic contrast, let us briefly rehearse the orthodox answer to this question. It would put the emphasis upon *epistemic,* rather than *agency,* and urge the point that only minds can *think* and hence engage in transactions and interactions with epistemic import. Indeed, the essence of mind is being a *res cogitans* (a thinking entity); only a *res cogitans* can be an *agens cogitans* (a thinking and acting entity). In a more recent and less metaphysically loaded version, the idea is that only beings with *representational* powers can be said to have knowledge. Man-made measuring devices can be keyed to certain parts of the world in ways that make them gauges with respect to that reality: they are useful to human beings in their cognitive exploration of the world. However, they are not themselves representational systems and, hence, not epistemic subjects.

Goldberg does not take this route. In fact, he allows that such non-minded things as measuring instruments may be said to *represent* the world.[11] Instead, he puts focus upon "agent" rather than "epistemic" in the term "epistemic agent," thereby shifting attention away from knowledge as a passive state of knowers towards what we *do* with it, particularly in interpersonal uses. With this move, a whole cluster of concepts standardly deployed in the description and assessment of human conduct also become relevant within the epistemic realm: duties, rights, and responsibilities. Social epistemology turns out to be a chapter of ethics, law, or even political philosophy. Although, in a derived sense, a person has responsibilities towards herself, the full sense of the notion applies only in interpersonal contexts. Responsibility is something one incurs in relation to others.

We also see a rationale for making minds central. Only minded things can owe—and be owed—duties, whether epistemic or other. This is a highly intuitive principle and Goldberg does not argue for it in his text. Indeed, it might be considered a rock-bottom intuition, neither needing nor allowing of supporting argument. As it happens, however, Goldberg has actually supplied such reasoning elsewhere; since this promises a deeper insight into the nature of epistemic agency and its link with mindedness, we should look at it.

The key texts are the monograph, "*Relying on Others: An Essay in Epistemology,*"[12] and the follow-up article, "Epistemic Extendedness, Testimony and the Epistemology of Instrument-Based Belief."[13] In the former, Goldberg demonstrates that epistemic assessment can be extended beyond the single individual. In the later article, he formulates this result in what he calls the Generic Epistemic Extended Mind Hypothesis (GEEM):

> For at least some cases in which a subject S believes that P, a proper epistemic assessment of S's belief requires an epistemic assessment of information processing that takes place in the subject's environment.[14]

In *Relying on Others*, Goldberg convincingly defends this principle, using as his primary example situations where S bases her beliefs on another person's sincere testimony. Now, however, a question arises. Why cannot GEEM be extended, in the spirit of externalism, to readings from thermometers, FMR scanners, and other such (non-minded) pieces as measuring apparatus?

In the article, Goldberg is concerned to show that this move can be—and should be—resisted. The position established in *Relying on Others* is that there are cases where the epistemic assessment of A's beliefs depends on an assessment of some other person B's beliefs. Another way of putting this is to say that A's cognitive processing is *extended* to encompass B's cognitive processing with respect to those beliefs.

Now, according to Goldberg, trouble ensues if we adopt the same policy when the role of person B is performed by a non-minded source of data, such as a measuring apparatus. Notice first that if we drop the requirement of mindedness, basically any law-governed natural process may be construed as somehow "measuring" or "recording" events in its vicinity, or even more global physical processes. For example, a tree may be construed as a chronometer, recording the passing of time by its production of annual rings. This is precisely the natural phenomenon featured in Goldberg's argument.[15] It involves a person S who is so disposed that whenever she observes a tree stump with n annual rings, she forms the spontaneous belief that the tree is n years old. S is completely ignorant, meanwhile, of the mechanism connecting the age of a tree with the number of rings and their resulting covariation; she just finds herself with such beliefs in the circumstances described, never questioning or doubting them.

Our intuitive reaction to this case, according to Goldberg, is to declare S epistemically irresponsible. The move from the observation that the number of rings on a tree stump is n to the belief that the tree is n years old is "the sort of transition that we expect a subject to support with evidence."[16] However, if we extend S's cognitive processing to include the mechanism whereby the tree produces an annual growth ring, the verdict must be the opposite. Now S will be considered epistemically responsible, as the situation now is similar to that in which a normal person trusts the deliverances of her eyes (in the absence of relevant defeaters). It is generally agreed that such a person commits no epistemic irresponsibility in doing so, even if she knows of no statistics about the reliability of human vision nor has any knowledge of the workings of the visual system that would imply its reliability.

Goldberg proceeds immediately to articulate the general principle of which this intuitive asymmetry is a symptom:

> [The] lesson is obvious: assessments of justification are normative, and insofar as information-processing operations are going to be relevant to such assessments, they ought to be operations which themselves are properly subject to the full range of normative assessments that go into epistemic assessment—including assessments of rationality and responsibility.[17]

This principle is indeed intuitively plausible, yet we may still feel that we have not been given a rationale for it. What has been shown is merely that our intuition with respect to the tree stump scenario is linked with further intuitions in a way that gives us confidence in its robustness; however, it has hardly been explained. We still want to ask, "What are the features of minds that warrant the attitude that epistemic responsibility can only be owed *by*, and owed *to*, precisely *minded* entities?"

There is a hint at a possible answer in Goldberg's article. The fundamental atom of all epistemic interaction is the *speech act of assertion* (chapter 1, section 2). Its most elementary occurrence is when one person tells another person something (individual testimony). More complex and institutionally regimented forms are found, for example, in written expert reports, scientific articles, and legal briefs. In uttering even the most humble assertion, a speaker incurs an epistemic obligation concerning concepts such as truthfulness or justification. We see that all more sophisticated cases of communication build and elaborate on this foundation.

This suggests that the "mindedness" of genuine epistemic subjects is constituted by the kind of intentionality and responsiveness to norms that is minimally required for engaging in linguistic interaction. This is a modernized version of a thought as old as philosophy itself; that is, to conceive of the "mind" or "reason" that distinguishes us as humans as being linked with our ability to use language.

This, however, raises the question how we should deal with the hyper-powerful "super computers" of the future that will converse with people in natural language and offer them advice on matters that exceed the cognitive powers of human beings (for instance, predictions about long-term trends in the stock market). Could such computers be held responsible for economic losses or even for market crashes? If not, is this because they are too limited in their linguistic repertoire to count as full-fledged language users and hence as having minds? Or, is it rather because the notion of sanctions does not make sense with respect to a computer, thus rendering the notion of responsibility vacuous? Goldberg sidesteps this issue; he thereby may be said to skirt, in the final analysis, why precisely *minds* matter in social epistemology.[18]

There is a related issue more pressing than these futuristic scenarios, however. What is Goldberg's stance concerning the epistemic responsibilities of social institutions, such as courts of law, professional advisory boards, and academic institutions? Do they count as "minded?" If the answer is yes, is this solely by virtue of their members having minds? If so, does this commit the projected social epistemology to "methodological individualism" in its analysis of such institutions? Moreover, would this imply that only the individual members of such institutions can incur epistemic responsibility for verdicts passed or recommendations made, if they produce unfortunate consequences? This raises an issue explored in recent work in moral philosophy.[19] Sometimes, collective bodies will make decisions and issue recommendations that diverge from the opinion of each individual member of the body, due to quirks of the procedure through which individual recommendations are aggregated into a collective one (such quirks can be shown to be unavoidable).[20] In such cases, it would be awkward to hold the individual members of the collective body responsible for decisions that they actually opposed and against which they had voted. This provides a reason to say that the collective *as such* is responsible, lest we end up with a "responsibility deficit" where nobody can be held responsible for a disastrous result. However, given Goldberg's stress upon the role of *minds* in social epistemology, this leaves him with the problem of how to attribute "mindedness" to such collective entities.

4. GOLDBERG'S PROGRAM FOR SOCIAL EPISTEMOLOGY

Having established (in *Relying on Others*) the legitimacy of the notion of distributed epistemic responsibility, Goldberg goes on to sketch out a research program for social epistemology. The overall strategy has been suggested already. The starting point is the basic component of epistemic interaction, the simple act of assertion whereby an individual speaker relays information to an interlocutor. Social epistemologists would then gradually draw ever-wider circles within the social sphere and analyze both the transformations and complications of epistemic agency that follow with each expansion and the changes of empirical methodology compelled by the wider scope.

They would also investigate the shifts in epistemic norms consequent upon such expansion of focus. The epistemic rules governing the simple situation of person-to-person testimony are explored in speech act theory. However, information passed on in this way may have repercussions for a much broader group of people than the participants in a conversation; such effects are subject to general moral assessment (think of the effects on a person when slander is passed around behind her back, or the effects of "hate speech" upon a social group).

A more complex issue still, and a favorite among adherents of ASE, is the societal role of experts. However, where ASE so far primarily has been concerned with the non-expert's epistemic predicament in deciding which experts to trust, Goldberg focuses on the experts' professional responsibilities in passing on information to lay people—or indeed in society's responsibility in instituting certain bodies of experts to serve, for example, on committees.

Goldberg suggests an over-arching normative framework for all these social practices. Knowledge sharing is a highly beneficial social institution, being a kind of epistemic division of labor, which, like division of labor in non-cognitive lines of work, brings huge gains in "productivity." And although Goldberg shuns such crude economic terms, it is clear that the underlying argument is a utilitarian one, where the goal is gains in general welfare.

The suggested social epistemology will be a naturalistic enterprise, drawing upon a broad range of empirical disciplines. Goldberg insists, however, on the autonomous normativity of social epistemology, like the classical epistemology from which it springs. Social epistemology is not committed to endorsing the norms inherent in any actual practice under investigation. Not only is social epistemology of Goldberg's variety reformatory of any given practice in the sense of comparing it critically with others, it is also prepared to adopt norms of assessment of its own. For instance, it typically would be critical of the various epistemic biases inherent in discriminatory social practices, be they sexist, racist, or others.

5. STEVE FULLER'S VERSION OF SOCIAL EPISTEMOLOGY

This is an appropriate point at which to turn to Fuller's article, since Fuller starts where Goldberg leaves off. As Fuller himself puts it in the paper, he likes to start *in medias res*, in actually existing situations, where epistemic responsibility is being shared, without a preparatory argument to justify this point of departure. In particular, he sees no need for extended argument to justify the very step from individualist to collective epistemology in the first place.

In the opening section, Fuller reaffirms two points on which he sees eye to eye with Goldberg and ASE. His version of social epistemology is *naturalistic*, looking to empirical research for an analysis of our communal cognitive practices. It is also *normative*, since it does not restrict itself to description of extant practices, but also considers potentially superior alternatives. Fuller defines normativity in the following manner: "For me a 'normative' approach is one committed to organizing the means available to bring about or maintain some desirable state-of-affairs" (chapter 2, section 1). We notice a strong affinity between this definition and a standard definition of economics as the "rational allocation of scarce resources to achieve prioritized goals."

Fuller continues to characterize his approach to social epistemology in terms that have close analogs in economics: "[We] need to invert the analytic epistemological understanding of the relationship of 'belief' and 'knowledge,' treating 'belief' less as something grounded in the past experience . . . than a risky projection into the future. . . ." (chapter 2, section 1). In economic terms, this would correspond to the difference between funds placed in passive and riskless investment, versus money invested in projects with great risks but also with prospects of big gains. We should note that this inversion was already made by CSE's Popperian ancestors in the philosophy of science. This is the conception of scientific theorizing as a matter of making "bold conjectures" (that is, epistemically risky hypotheses) that are not guaranteed by induction on past experience, but are rather pursued for their promise of future scientific gains.

To Fuller, it is a truism that social epistemology is about the distribution of a particular kind of basic human good; that is, knowledge. Moreover, we are talking about distribution on a large social scale, not on the micro-scale of the speaker-listener dyad examined by speech act theory. Fuller operates on a societal macro-scale, where such normative concerns become a matter to be handled by social institutions.

Fuller chooses the institution of law as his example, and we are not surprised to find that he takes the Law and Economics literature as his point of departure. A classical issue in this literature is the Polluter vs. Resident problem, which concerns the legal adjudication of situations where one party's actions—for instance, the production activities of a chemical business—encroach upon another party's enjoyment of their property, for example by releasing toxic fumes into a residential neighborhood. If property rights are considered inviolable, the legal verdict would most likely come out in favor of the residents and the business would be ordered to restrict its operations—perhaps at the price of a considerable reduction in general welfare, in the form of economical cutbacks, lost jobs, and other consequences.

Fuller examines an alternative legal construal; that is, the liability approach proposed by Calabresi and Melamed,[21] according to which the matter should be handled through an economical compensation to the injured party. Moreover, they propose a symmetrical analysis of the situation. The residents could be considered the injured party, to be economically compensated for the reduced benefit from their gardens, the decreased property value, and so on. On the other hand, the business might be construed as a party that has suffered a loss of freedom to use its land and production facilities as it pleases, and thus should be compensated accordingly. On which side the decision falls should be largely determined by considerations of which policy would produce the most general welfare.

According to Fuller, a similar policy should be adopted in social epistemology in general. When investing in major scientific-technological-administrative projects, the theoretical outcomes (or practical consequences) of which are uncertain, as they will be if the projects are truly innovative, we should proceed in the face of the potential hazards, recognizing that the negative effects could impact unevenly upon people. We should proceed with a view to the greater benefits for the total population—and with a commitment to compensate the injured parties for their sufferings.

6. FULLER ON EPISTEMIC AGENCY

Next, Fuller proceeds to draw some general implications from this case for the notion of an agent *per se*, starting out from the familiar idea that agency implies responsibility, which in its turn presupposes ability. Kant articulated the relationship between ethical responsibility and ability in the famous slogan, "ought implies can." This slogan can be read in two directions, however: either (from left to right) taking obligations as given and inferring capacities from this, or (from right to left) by taking human powers and their limitations as given and deducing constraints upon moral obligations from this.

Fuller introduces a third reading of the principle, however, which forms a sort of compromise between the first two. This is to the effect that the fundamental duty is that of *transforming oneself* into a kind of being that is capable of living up to even the strictest moral imperatives. Fuller suggests that the liability model of responsibility will serve to enhance agency in this sense: By allowing other agents to engage in activities that involve risks to ourselves (but risks that will be compensated), we may derive benefits that will give us larger freedom of action in the future, by way of a more comprehensive set of options.

Fuller understands this proliferation of options in a very radical sense. It is not only a matter of the options that depend on external resources, material or social, but our "internal" options as well. We can proliferate these by transforming ourselves into superior human beings. We do this by drawing on sophisticated biological knowledge, which even will enable us to manipulate the human genome. When properly deployed, this knowledge may literally make us morally superior persons—by suppressing, for instance, the human genes that code for selfish conduct detrimental to the pursuit of social goals, or those genes that cause us to pursue small promixate rewards and neglect much larger, but more distant, ones.

This is the point at which Fuller's reflections on Calabresi and Melamed's legal innovations connect with the general issue of human agency. In looking to science for ways to improve the human species as a moral agent (for instance by manipulating the human genome), we must drop the risk-averse attitude characteristic of traditional epistemology (and of inductivist, pre-Popperian philosophy of science). We must be willing to engage in risky, uncertain scientific projects that will not necessarily benefit people who are currently alive, but may bring great benefits to future generations. Fuller urges us to adopt what Max More has dubbed the *proactionary principle*, in place of the *precautionary* principle, which supposedly rules current science policy.[22] The latter would ban scientific innovations, even those of great promise, if they involved minimal (but not precisely calculable) dangers, such as may be incurred in projects about genetic modification. The proactionary principle, on the other hand, would go for the probable gains in utility, ignoring the minuscule risks of disaster.

Fuller does not take the matter further in the present text, but in other writings, notably *Preparing for Life in Humanity 2.0*, he has explored the issue in considerable detail.[23] He envisages—and welcomes—a future process of human genetic modification that will eventually produce a completely new species, "Homo 2.0," to replace Homo sapiens. Fuller thus joins up with the group of modern intellectuals who have heralded the coming of a "posthumanist" era. To Fuller, however, this would merely spell the end of one particular humanoid species (Homo sapiens), but would not spell the end of humanity, since the replacement would be another humanoid species. Fuller would probably even argue that this replacement would be in accordance with a defining trait of Homo sapiens as a humanoid species: that is, its constant striving for self-improvement and self-perfection. Fuller points out that, in the Western tradition, this perfectibility has always been seen as a characteristic of man, whether it was conceived in religious terms (man seeking redemption from sin) or in secular, cultural terms (man cultivating his rational side and overcoming his animal passions).

There is another aspect of Fuller's conception of agency that may be gleaned from his writings on the successor of *Homo sapiens:* the ultimate source of the obligations that (epistemic) agents incur towards one another. Fuller's stance is suggested by his endorsement of the proactionary principle. This is fundamentally a *utilitarian* principle, which is fully consonant with Fuller's analysis of legal norms, as examined above. Their basis is fundamentally utilitarian; it is all a matter of creating general welfare. Incidentally, Fuller himself refers to his ethical stance as a "superutilitarianism," which is distinguished from standard utilitarianism on two points. First, it is risk-seeking, being willing to engage in projects that promise huge gains, but that also involve considerable (but not precisely calculable) risks. This policy of course follows from Fuller's general epistemic stance. Second, "superutilitarianism" adopts a very long time perspective, where the welfare of current generations might be sacrificed for that of future ones.[24]

Does Fuller's notion of agency answer the question that exercised us in connection with Goldberg; that is, the significance of "mindedness" (consciousness) in (epistemic) agency? An answer might be thought to follow from Fuller's utilitarian construal of social obligation. For obligations to exist, the notion of sanctions must apply; this, in utilitarian ethics, means the deliberate infliction of a utility loss upon agents. Non-minded entities, such as computers, do not have utilities or disutilities; hence, they cannot incur obligations (nor do we have obligations towards them). We conclude that they do not count as (epistemic) agents.

Fuller has not addressed such issues; hence, there is no point in pursuing them here. Nor is there any indication, in the text before us, of Fuller's stance toward attributing responsibility to collectives per se, rather than to their individual members. We may note, however, that on utilitarian ethics such as Fuller's, businesses and other collectivities may, with perhaps a slight extension of terms, be said to be punishable (for example, by fines imposed on the firm, not upon its individual representatives). Hence, Fuller's utilitarian construal of responsibility would seem to make room for considering businesses and similar incorporated collectives as agents.

7. IN CONCLUSION

By way of a brief summary: The two articles we have discussed illustrate characteristic features of ASE and CSE, respectively. ASE has been much concerned with securing the legitimacy of the very move from individual to social epistemology; the shadow of Cartesian epistemology still falls upon it. Hence, to Goldberg, the territories to be explored, once the social turn has been made, are still largely distant vistas. As a result, his article is highly programmatic—as its title, indeed, says. By contrast, Fuller takes the social turn for granted, and is already busily at work implementing the research program heralded by it. He takes law as a concrete case, going into considerable detail to explore current legal theory as a source of insights for social epistemology.

If we put the two contributions end to end, what results is a fairly continuous argument, which starts out by establishing the soundness of the fundamental step from individual to social epistemology and ends with a concrete proposal for how the latter, with its normative goals and naturalistic methodology, might engage with such societal institutions as civil law. I hasten to add that both authors would no doubt have reservations about many (but different) details of the resulting combined argument. Still, the possibility of a fairly seamless fusion of the two contributions suggests that the opposition between the two underlying approaches (ASE and CSE) is more apparent than real.

Along the way, both authors offer inputs to a notion of an epistemic agent suitable for social epistemology. Still, they leave important lacunae, both concerning the distinction between "minded agent" and "non-minded machine" and regarding the status of collective agents. Ultimately, both Goldberg and Fuller are more interested in exploring the new territory that has opened up after the social and naturalistic turn in epistemology than in plumbing the metaphysical depths of agency.

NOTES

1. Karl R. Popper, *Conjectures and Refutations* (London: Routledge & Kegan Paul, 1963).

2. Thomas S. Kuhn, *The Structure of Scientific Revolutions* (Chicago: University of Chicago Press, 1963, 2. ed. 1970).

3. Cf. David Bloor, *Knowledge and Social Imagery* (London: Routledge & Kegan Paul, 1976); Harry S. Collins, *Changing Order* (Chicago: University of Chicago Press, 1985/1992).

4. Steve Fuller, *Social Epistemology* (Bloomington: Indiana University Press, 1988, 2. ed. 2002).

5. Alvin I. Goldman, 1986. *Epistemology and Cognition* (Cambridge,MA: Harvard University Press, 1986).

6. Alvin I. Goldman, *Knowledge in a Social World* (Oxford: Oxford University Press, 1999).

7. Bruno Latour, *The Pasteurization of France* (Cambridge, MA: Harvard University Press, 1988).
8. Bruno Latour, *We Have Never Been Modern* (New York: Harvester Wheatsheaf, 1993).
9. Bruno Latour, *The Pasteurization of France*.
10. Andy Clark and David Chalmers, "The extended mind." *Analysis* 58, 1 (1998): 7–19.
11. Sanford C. Goldberg, (2012). "Epistemic extendedness, testimony, and the epistemology of instrument-based belief." *Philosophical Explorations*, Vol. 15, No. 2, (2012): 181–97. (Goldberg 2012, p. 192).
12. Sanford C. Goldberg, *Relying on Others: An Essay in Epistemology* (Oxford Scholarship Online, 2010).
13. Sanford C. Goldberg, "Epistemic extendedness, testimony, and the epistemology of instrument-based belief." *Philosophical Explorations*, Vol. 15, No. 2, (2012): 181–97.
14. Goldberg, "Epistemic extendedness": 181.
15. Ibid., 189.
16. Ibid., 189.
17. Ibid., 190.
18. Ibid., 194.
19. List, Christian List and Philip Pettit, *Group Agency: The Possibility, Design, and Status of Corporate Agents* (Oxford: Oxford University Press, 2011).
20. List, Christian List and Philip Pettit, Philip, "The Aggregation of Sets of Judgments: An Impossibility Result." *Economics and Philosophy,* 18 (2002): 89–110.
21. Guido Calabresi and A. Douglas Malamed, A. "Property Rules, Liability Rules, and Inalienability." *Harvard Law Review* 85. (1972): 1089 ff.
22. Steve Fuller, S. *Preparing for Life in Humanity 2.0.* (London: Palgrave Macmillan (2012): 27.
23. Ibid.
24. Ibid., pp. 63–65, p. 81.

REFERENCES

Bloor, David. *Knowledge and Social Imagery*. London: Routledge & Kegan Paul, 1976, 2. ed. Chicago: University of Chicago Press, 1991.
Calabresi, Guido & Malamed, A. Douglas. "Property Rules, Liability Rules, and Inalienability." *Harvard Law Review* 85 (1972): 1089 ff.
Clark, Andy & Chalmers, D. "The extended mind." *Analysis* 58, 1, (1998): 7–19.
Collins, Harry S. *Changing Order*. Chicago: University of Chicago Press, 1985, 2. ed.1992.
Fuller, Steve. *Social Epistemology*. Bloomington: Indiana University Press, 1988, 2. ed. 2002.
Fuller, Steve. *Preparing for Life in Humanity 2.0*. London: Palgrave Macmillan, 2012.
Goldberg, Sanford C. *Relying on Others: An Essay in Epistemology*. Oxford Scholarship Online, 2010.
Goldberg, Sanford C. "Epistemic extendedness, testimony, and the epistemology of instrument-based belief." *Philosophical Explorations* 15, No. 2 (2012):181–97.
Goldman, Alvin I. *Epistemology and Cognition*. Cambridge, MA: Harvard University Press, 1986.
Goldman, A.I. *Knowledge in a Social World*. Oxford: Oxford University Press, 1999.
Kuhn, Thomas S. *The Structure of Scientific Revolutions*. Chicago: University of Chicago Press, 1963, 2. ed. 1970.
Latour, Bruno. *The Pasteurization of France*. Cambridge, MA: Harvard University Press, 1988.
Latour, Bruno. *We Have Never Been Modern*. New York: Harvester Wheatsheaf, 1993.

List, Christian & Pettit, Philip (2002): "The Aggregation of Sets of Judgments: An Impossibility Result." *Economics and Philosophy*, 18 (2002): 89–110.
List, Christian & Pettit, Philip (2011): *Group Agency: The Possibility, Design, and Status of Corporate Agents*, Oxford: Oxford University Press, 2011.
Popper, Karl R. *Conjectures and Refutations*. London: Routledge & Kegan Paul, 1963.

FOUR
Fuller's Social Epistemology and Epistemic Agency

Francis Remedios and Valentine Dusek

Fuller's unique version of epistemic agency is social constructivist with a truth-oriented standard by which the agent is evaluated.[1] Fuller's version of epistemic agency is in opposition to ASE's (i.e., analytic social epistemology) version of epistemic agency, which is doxastic (i.e., belief) and truth-oriented. The advantage of Fuller's version to ASE's is that it is ontologically open to different bearers of epistemic agency such as persons, corporations, and universities while ASE's is ontologically restricted to humans only. Furthermore, Fuller's social epistemology accounts for epistemic change of the epistemic agent based on changes in boundary conditions due to converging technologies of biotechnology, genetic engineering, and synthetic biology to transhumanism. ASE does not have an account of these types of changes to the epistemic agent, because ASE is not interested in social and scientific changes which may transform the epistemic agent.

We start with a discussion comparing Fuller's social epistemology to ASE on how knowledge should be organized. Connected to Fuller's version of transhumanism are his notions of personhood and epistemic agency. For instance, if it is the case that human beings are becoming humanity 2.0 (i.e., radically outstripping its previous boundaries), Fuller must rework pre-existing notions of personhood and epistemic agency to accommodate his transhumanism.

Fuller emphasizes agent-oriented epistemology over object-oriented epistemology, which is aligned with ASE. We contrast agent-oriented and object-oriented social epistemologies. In the same vein, we discuss expertise to contrast agent-oriented social epistemology, which does not defer uncritically to experts, and object-oriented social epistemology, which does so with respect to domains of knowledge. We proceed with a discussion of Fuller's version of social epistemology as a version of cognitive economics, as it is the epistemic agent who gains knowledge and leverages beliefs into action. We close with a discussion of Fuller's conception of thick and thin conceptions of agency.

1. FULLER'S SOCIAL EPISTEMOLOGY AND ANALYTIC SOCIAL EPISTEMOLOGY

Fuller's version of social epistemology is significantly different from ASE. First, his version is not solely concerned with truth, whereas ASE tends to analyze the social dimensions of knowledge for the sole purpose of assessing truth claims. In this regard, ASE's objective is often restricted to the evaluation and promotion of true belief. The distinguishing feature of Fuller's approach is his additional concern for extra epistemic considerations that far exceed the assessment of truth claims. For instance, Fuller's social epistemology 1) addresses humanity's search for transcendence, in which humans can become more than they presently are through science and technological advancements, 2) seeks to *advise* on *policy* issues in order to increase knowledge production, and 3) addresses the problem of a unified theory of everything, which he regards as the aim of science and requiring a systematic representation of reality.

It is additionally important to note the different approach that Fuller takes towards knowledge from that of ASE. ASE focuses on the individual's or group's beliefs and knowledge that permit them to interact with the world. Following Continental epistemology, Fuller's social epistemology is focused on the collective memory of scientific institutions and inquiry's historical processes that illuminate epistemological problems. Like Popper on the falsification of scientific theories, Fuller believes that good scientists fight with their collective memory of science and their awareness of its many errors.

Due to his above views on knowledge, Fuller additionally considers epistemology to be a style of metaphysics in which the epistemic agent constitutes or constructs reality as part of the epistemic process. This opposes the belief that reality is fixed.[2] Fuller is following Hegel whose phenomenology of mind is an epistemology which transcends subject and object to an idealist metaphysic, that is the world is not portrayed as totally mind independent but partly constituted by the mind. For Fuller (and his constructivist reading of Hegel), the epistemic agent creates the conditions that render knowledge possible and establishes what it means to possess knowledge. Counter to Hegel and Fuller is the classical realist's correspondence theory of truth, in which the epistemic agent is not important because knowledge is a state that mirrors reality.

2. FULLER'S HUMANITY 2.0, CHANGES AND EXPANSIONS OF PERSONHOOD AND AGENCY

Fuller has a wide notion of personhood compared to the ASE notion of personhood, which tends to include only humans. With humanity 2.0, Fuller views the ontology of person to be open, because a "person" can refer to a human being as well as non-human entities such as corporations, nation states, and universities. What they have in common is that they are all producers of knowledge. He considers non-human entities such as corporations, nation states, and universities, to be 'persons' because they have the capacity to act.

Current or future advancements in biotechnology, genetic engineering and the even more extreme modifications made possible by "synthetic biology" (the implanting of whole genomes from one organism into another) raise the question of what it means to be human, and of what the distinctiveness of humanity is. Examples from biotechnology include humans who are becoming cyborgs with computer chips implanted in bodies and synthetic biology where Chinese scientists are editing genes of human embryos. Fuller's transhumanism is the view that humanity can be enhanced or redesigned through technology, e.g., prosthetic devices, "smart drugs" that improve intelligence (and other mental capacities), and implanted computer chips. Similarly, converging technologies, such as biotechnology, nanotechnology and computer technology, are also transforming and enhancing humanity—humanity 2.0.[3]

In law, there is not a lot of ontological depth to corporate agents. Likewise, Fuller's notion of person is based in the legal definition of 'liability' in which a 'person' (as either an individual or corporation) is held responsible for her or his actions.[4] STS (i.e., Science, Technology, and Society) has transferred responsibility from corporations, which are legal persons, to networks where responsibility is distributed across the network. For example, in actor network theory, both humans and non-humans as parts of a network possess the capacity to act.[5]

For Fuller, epistemic agency, which includes humans and non-human entities such as corporations, nation-states, and universities, is a form of personhood. Though the epistemic agent is socially constructed, the standard by which the epistemic agent is evaluated is truth-oriented. For Fuller, truth is a systematic representation of reality, a grand unified theory of everything. To achieve this type of scientific knowledge, it is an open question as to the type of agent that would be most appropriate. It would not necessarily be an individual scientist such as Einstein or Hawking—whom some may consider a cyborg. A 'person' may also be a computer or a group of scientists.[6] For Fuller, humanity could continue even if homo sapiens end.

A major difference between Fuller and ASE concerns epistemic agency as non-human epistemic system. In opposition to this view, Goldman states only knowers, who are humans that have beliefs, are epistemic agents.[7] This is ontologically closed, in that it does not permit anything else to count as an agent. In Fuller's more ontologically open view, any epistemic system can be an epistemic agent.

3. AGENT-ORIENTED VS OBJECT-ORIENTED SOCIAL EPISTEMOLOGIES

Fuller distinguishes between *agent-oriented*, which concern the epistemic agent who constructs knowledge, and *object-oriented* social epistemologies, which concern the nature of knowledge as divided and based on its objects. For agent-oriented social epistemology, the knower unifies knowledge based on a personal world view, like a medieval master of liberal arts such as Goethe. Fuller is sympathetic to German idealism in which the knowledge process is incorporated into the agent. The ideal end of this process is an epistemic state in which lies the achievement of the unification of everything in the agent's mind in the way of Goethe and Hegel.[8]

Fuller's agent-oriented social epistemologist develops knowledge policy according to the normative governance of science. Fuller favors agent-oriented social epistemology over object-oriented social epistemology, because he believes social epistemology ought to concern how knowledge can be pursued normatively.[9] ASE tends more towards object-oriented social epistemology in which the epistemic agent relies on experts in discrete fields of study.[10]

Agent and object-oriented social epistemologies are ideal types. The distinction between the two ideal types may be exaggerated to highlight the differences between them on normative implications in the history of philosophy. For example, agent-oriented social epistemology is characterized as idealist/rationalist, and object-oriented social epistemology is characterized as empiricist, in that agent-oriented social epistemologists believe that the agent's mind constructs knowledge and object-oriented social epistemologist believes that through observation the world divulges itself. Agent and object-oriented social epistemology can be considered as two ends of a continuum in the same way that individual and collective epistemologies can be considered as opposite ends of a continuum.

4. EXPERTISE

We live in a world in which expertise is highly valued in science, the justice system, universities and among professionals in Western influenced societies. On expertise, the normative implications are different between object-oriented and agent-oriented social epistemology. For object-oriented social epistemology, there are strong normative implications with knowledge divided into domains with epistemic agents who rely on experts in each of those domains. For object-oriented social epistemology, the objects of knowledge dictate what needs to be known, in turn, epistemic agents rely on experts who analyze those domains and authorize epistemic goals. ASE emphasizes the virtues of knowledge expertise with trust and deference to authority. Agent-oriented social epistemology is anti-expert, in which the epistemic agent seeks to integrate different domains of knowledge into her or his own aims and goals. With different normative implications for the agent-oriented social epistemologist, Fuller recommends critical purport of experts. Just as the utopian socialists and the early Marx criticized the division of labor in economics, so Fuller's notion of the epistemic agent rejects the uncritical acceptance of the division of intellectual labor. This view contrasts with object-oriented social epistemology in which the expert is relied on to dictate the ends of knowledge. For instance, Catherine Elgin's epistemic agent uses Hillary Putnam's notions of "division of intellectual labor" and obedience to (epistemic) authority—the trusting of experts.

There are many cases where excessive reliance on the division of intellectual labor and trust in experts of object-oriented social epistemology seem inappropriate. In recent years both social scientists and lay people have become more aware of the economic and political issues that have affected the deliverances of scientific experts or specialists in various fields. Some of the most egregious cases have been in testing, where prestigious MDs have accepted pay for signing their names to favorable articles actually written by employees of the pharmaceutical industry hoping to market the drug. Less egregious, but still misleading, are articles favoring the safety of a drug authored by MDs with an interest in marketing that drug.[11] Fudging and massaging of data also occur in military weapons testing[12] and in the process of peer review, grant awards, and other support of publications favoring politically influential scientists' research programs.[13] "Experts" with conflicts of interest involving their paymasters or their own research agendas are, increasingly often, not expert sources of knowledge. The epistemological agent therefore needs to participate in the evaluation of the material provided by the alleged expert.

It is not so easy to separate "the scientists" pronouncements on their specialties from the more ideological, religious (or anti-religious) or policy-driven pronouncements. For instance, Edward Teller was undoubtedly the authority on the physics of hydrogen bombs (after all he invented them), but his claims about the benefits of nuclear testing, for instance, though invested with authority, were not part of the "core" science. Linus Pauling, similarly, used his expertise in quantum and organic chemistry to deny the safety of nuclear testing, emphasizing its effects on the body.[14]

A particularly egregious misuse of supposed expertise is when some scientists and philosophers use their prestige as experts on biology to pronounce their religious or non-religious views. The "new atheists" such as Richard Dawkins claim that atheistic conclusions derive from Darwinian biology. Michael Ruse outraged Dennett by saying that by teaching atheism as part of Darwinist science "the new atheists" are liable to be faulted for teaching religion (atheism) in public schools. Dennett replied by excommunicating Ruse from the community of respectable philosophers of biology. Dawkins, Dennett, and the historian of population genetics Provine claim that their assertion of the truth of atheism is part of their science, while others would strongly disagree and claim the "new atheists" are simply foisting their (non-)religious beliefs as science.[15]

"Obedience to (epistemic) authority" of the "experts" is not always the most successful policy of knowledge acquisition. For instance, The Electric Power Institute is heavily financed by the electric power industry. It does risk benefit analyses of issues concerning the effects of proximity on the environment and human neighborhoods near electric power plants, almost always downplaying the risks. Similarly the scientists financed by the chemical industry, pesticide industry, and food processing industry and University Agriculture departments are largely financed by the food industry, so reports are often highly biased toward support of the funding source.

Ames, chemical expert, inventor of the Ames Test, a microbiological test, was famous for his debunking of claims about carcinogenic potential of various newly synthesized organic chemicals. His test was widely reported debunking of the carcinogenic potential of the pesticide Alar, commonly used on apples. Newspaper editorials ridiculed fear of Alar, one citing a street person dumpster diver who refused an apple offered to him because of his fear of Alar. It later turned out that Alar is carcinogenic. Even if Ames had no direct support from the chemical industry, he was motivated to defend synthetic organic chemistry.[16] However, newspapers continued to use Alar as an example of ridiculous yuppie ecology nonsense long after the real carcinogenic effects were documented.

Many scientists have debunked and ridiculed various informal claims of cancer among children; so-called cancer clusters, including leukemia; and claims of local residents citing massive pollution as a cause. Yet, the high rate of cancer at Love Canal was first largely brought to the general public's attention, not by scientific experts, but by housewives.[17]

Famously, AIDS patient activists moved the medical profession to dispense with double blind tests when testing potential AIDS drugs. Again the "experts" opposed the testing without control groups, when in fact the alternative for those denied the drugs was death. ACT-UP and other AIDS activists succeeded in changing the way medical research was done.[18] Similarly women's health activists changed the procedures of surveys and sampling with respect to diseases of women, notably by including more women in samples, showing a much higher frequency of cardiovascular disease in women than had been assumed when samples included almost solely males.[19]

Another area in which deference to expertise and specialists is undesirable is in hereditary IQ studies. The "experts" in behavioral genetics and IQ studies are almost all strongly committed to hereditarianism. The critics of hereditarian theories of IQ are people such as Noam Chomsky, Steve Gould, Richard Lewontin, David Layzer, and Ned Block.[20] This group includes a linguist, two evolutionary biologists, a physicist, and a philosopher. None of these brilliant and mostly highly mathematically adept are members of the professional community of IQ heredity scholars, who almost uniformly support hereditarianism and in a not unsubstantial number of cases advocated so-called "race realism," or what used to be called "scientific racism." The mostly hereditarian "professionals" dismiss critics as not being "specialists," credentialed or experts in the field.[21]

5. COGNITIVE ECONOMICS

Following Nicholas Rescher [22] and Charles S. Peirce,[23] Fuller sees social epistemology as a version of cognitive economics that concerns the economics of knowledge. Fuller's cognitive economics additionally concerns knowledge policy, which addresses the manner in which knowledge ought to be organized.

Fuller distinguishes between demand-side and supply-side epistemologies. James's "Will to Believe" is an exemplar of demand-side epistemology, in which the epistemic agent voluntarily leverages belief into action and thereby gains knowledge. W. K. Clifford's epistemology is an exemplar of supply-side epistemology, in which epistemic agents protect themselves from an uncertain future by placing their knowledge on a foundation. [24] Fuller practices demand-side epistemology, which is proactionary, in that it seeks growth and prosperity through risk taking. This view opposes supply-side epistemology, which is precautionary and thus avoids risk taking.

Fuller's example of the precautionary side is Elgin's notion of an epistemic agent who relies on experts to set authority on a domain of knowledge and then believes the conclusions of the expert unless the agent has independent knowledge to the contrary.[25] Elgin's notion of belief is suspect, because based on a different interpretation of evidence, the agent can reject the conclusion of the expert and because having a belief does not imply that the agent should act on it. Fuller's view is that the agent should take on information and then decide whether to believe that information and to act on it.

Fuller's preferred epistemic agent follows proactive rather than reactive or precautionary principles. This leads her or him to advocate greater risks in scientific research and the encouragement of arrangements whereby people can knowingly volunteer as subjects of risky or dangerous experiments. He, for example, supports voluntary submission to risky brain modifications as seen in his review of "Critical Neuroscience,"[26] Jose Delgado's experiments on brain implants and brain control,[27] and Walter Penfield's operations[28] to remove parts of the brain in epileptic patients. He, however, condemns forced participation as seen in Nazi experiments on concentration camp prisoners and the American Tuskegee Experiment of letting syphilitic African Americans die when cures for the disease became available.

Though Fuller refers to Milgram with respect of "empathetic cruelty," it is surprising that sociologist Fuller does not discuss the proactive nature of social science experiments such as Zimbardo's "Stanford Prison Experiment" or Milgram's "Obedience to Authority" experiment.[29] (Oddly, Milgram and Zimbardo were professors at Yale in 1960.) A major reaction to these experiments among social psychologists was not to wonder what the experiments revealed, but to provide ethical objections to the experiments. Nonetheless, Zimbardo's and Milgram's social psychology experiments became widely popularized and shocked the public about the ease with which normal, average people could become sadists in the right social setting. These experiments explain how Nazi concentration camp guards and committers of war crimes in Vietnam and torturers at Abu Ghraib in Iraq need not be monsters and could be perfectly normal people before they were pu t into positions of extreme authority and/or made obedient to extreme authority.

The Nazi hypothermia experiments, despite the unwillingness of many to make use of the results of these murderous experiments, produced genuine data of use to preparing pilots who might be downed in arctic waters,[30] but many other Nazi "experiments" had no knowledge payoff. Thus, some sort of mechanism for the protection of human subjects should be in place, even if it is a much more liberal one than presently exists. Such a policy, while more attuned to the benefits of research on human subjects, should ensure that subjects willingly participate in projects involving risks and receive full indemnification if negative results occur.

6. CONCLUSION

Fuller's social epistemology has a number of advantages over analytic social epistemology. First, Fuller's approach considers normative results of knowledge other than truth. Second, Fuller's epistemology allows for changes in the knower such as those adumbrated by transhumanism rather than assuming a constant nature of the knower. Third, Fuller's account considers corporate and communal knowers in addition to individual knowers. Fourth, Fuller is more critical and reflective about obedience to alleged scientific expertise than its analytic alternative. Finally, Fuller's account encourages a proactive rather than a cautionary approach allowing more daring scientific investigation of human behavior and physiological structures.

Though we support these advantages of Fuller's notion of agency over ASE's notion of agency, we find fault with some of the aspects of Fuller's notion of agency that he attributes to scientists. We agree with Downes [31] that Fuller appears to apply a thin notion of agency to scientists. We note that Fuller's notion of "scientist," which lacks psychological richness and is associated with a thin notion of agency, seems to be incompatible with Kahneman's and Tversky's experimental results, [32] which appears to favor a thick notion of agency:

> They contend that humans do not reason according to the rules of probability and decision theory As a result of their experimental work, Kahneman and Tversky introduce agents who are psychologically rich. The agents' reasoning can be biased, and rather than being guided by rational choice theory, the agents' reasoning is guided by heuristics. [33]

Certainly there are a minority of working scientists who do reflect in a disciplined way concerning scientists' behavior. One can say that they are not philosophers or sociologists of science, but are reflecting in an informal and possibly philosophical manner on their activities. For example, there are scientists such as Paul Gross, Norman Levitt, and Alan Sokal who attacked the reflective discourse of sociologists and anthropologists. Other mathematicians and scientists such as Gabriel Stolzenberg, David Mermin, and Jay Labinger have reflected on the social nature of and aims of science, or, at least, their own science. For example, Jay Labinger wrote a critical study and history of the status of inorganic chemistry . [34]

Clearly some scientists, whatever one may think of their policy views, have also been knowledge policy analysts. Oppenheimer and Teller as dueling government advisors on nuclear weapons policy, James Watson as head of the Human Genome Project, Bernard Davis as the organizer of opposition to it, Teller and Pauling holding opposite views on the benefits and risks of nuclear testing, Warren Weaver playing a major role in the funding of molecular biology (and opposition to British Marxist and organismic biophysicists) [35] and Frederick Lindeman during World War

II. All these individuals were advisors on knowledge policy as well as sometime scientists.

Though Fuller has redefined his notion of a knowledge policy analyst to be an institution embodied in one person, a legislature or dispersed groups, [36] Fuller does not want ontological depth or thickness to the knowledge policy analyst. [37] We aver that the collective epistemic agent, who is no less than the individual epistemic agent, needs more "thickness" to be understood. Institutions have directives and missions that may include more than the purely epistemic and ethical directives of the knowledge policy analyst. Just as the individual agent may have not only normative epistemological demands on knowledge production, the collective agent may have such also. In this regard, institutions as well as individuals might be said to have "world-views" and general frameworks for evaluation of knowledge.

Here are just a few concluding examples of the manner in which the "world-views" of individuals and institutions can constitute frameworks for epistemic evaluation. Let us start with the radical Puritan scientific movement of the seventeenth century. The movement demanded knowledge that would have beneficial, practical, and positive results such as agriculture, medicine, and commerce. Baconian science policy heavily influenced the Puritan radical scientists, but religious conceptions of a return to an imagined harmonious Old Testament society and a Kingdom of God on earth ended Bacon's influence. This was, at least until, as Marx said, "Locke supplanted Habbakuk."[38] An extreme of this notion of the Old Testament is Francis Bacon's claim that technology and knowledge of nature will foster a return to the innocence and immortality of Adam and Eve before the Fall.[39] Scientific institutions that have campaigned for the elimination of creation science or Intelligent Design from secondary school biology lessons, whether one agrees with them or not, are also examples of institutions claiming to have a purely scientific agenda. Similarly, collective campaigns by mathematicians and physicists to keep social theorists of science out of the Institute for Advanced Study at Princeton and the forced retirement of an editor who disagrees with them were supposedly done to reject shoddy scholarship. However, they were more concerned with shoring up faith in scientists, increasing government funding, and suppressing new disciplines of thought critical of science (e.g., cultural and science studies).

NOTES

1. Remedios thanks Steve Fuller for comments on this paper. Remedios and Dusek thank Patrick Reider for comments on the paper.
2. Francis Remedios, "Review of Knowledge: The Philosophical Quest in History," *Metascience*, forthcoming.

3. Francis Remedios, "Knowing Humanity in the Social World: A Social Epistemology Collective Vision?" in *The Future of Social Epistemology*. ed. James Collier (London: Rowman & Littlefield International, 2015).

4. Steve Fuller, "A Sense of Epistemic Agency Fit for Social Epistemology," chapter 2, section 2. This is different from Wennemann's concept of moral personhood: Daryl Wennemann, *Posthuman Personhood*. (Lanham, MD: University Press of America, 2013).

5. Michel Callon, "Some Elements of a Sociology of Translation: Domestication of the Scallops and the Fishermen of St Brieuc Bay," in *Power, Action and Belief: A New Sociology of Knowledge*:196–233. ed. John Law (London: Routledge & Kegan Paul, 1986). Bruno Latour, *Science in Action, How to Follow Scientists and Engineers through Society* (Cambridge, MA: Harvard University Press, 1987). A neglected American philosopher, who developed at length a concept of corporate agency is Elijah Jordan, a kind of homegrown Indiana Hegel and/or Whitehead. Jordan may go too far in emphasizing corporate at the expense of individual agency, but he sees individual agency as issuing in large part from its relations to corporate bodies. Elijah Jordan, in *Forms of Individuality*, Indianapolis: Progress Publishing Co., 1927, surveyed and criticized concepts of individuality as difference, as content, and as intent, and supported a relational notion of individuality. His notion of agent is one in which corporations have a "will" (tendencies to growth and expansion) of their own, but lack an "intellect." It is the role of individuals to use their reason to supply the intellect to guide the corporation. Jordan's notion of corporate individuality and criticisms of other forms of individuality are explained and developed in Barnett, George, and Jack Otis. *Corporate Society and Education* (Ann Arbor: University of Michigan Press, 1961) and Glenn Negley, *The Organization of Knowledge* (New York: Prentice Hall 1942), 301–26.

6. Stephen Hawking warns artificial intelligence could end mankind http://www.bbc.com/news/technology-30290540.

7. Alvin Goldman. "Systems-Oriented Social Epistemology," *Oxford Studies in Epistemology*, Vol. 3: 189–214. ed. T. Gendler and J. Hawthorne (Oxford: Oxford University Press, 2010).

8. Remedios, "Review of Knowledge."

9. Steve Fuller, *Knowledge: The Philosophical Quest in History* (New York: Routledge, 2014), 14–15.

10. Remedios, "Review of Knowledge."

11. Sheldon Krimsky, *Science in the Private Interest: Has Corrupted Biomedical Research?* (Lanham, MD: Rowman & Littlefield), 2004.

12. Gary Taubes, "Postol versus the Pentagon," *Technology Review*. April, 2002. http://www.technologyreview.com/featuredstory/401412/postol-vs-the-pentagon/.

13. Robert Bell, *Impure Science: Fraud, Compromise, and Political Influence in Scientific Research* (New York: Wiley, 1992).

14. Melinda Gormley and Mellisae Fellet, "The Pauling-Teller Debate: A Tangle of Expertise and Values, *Issues in Science and Technology*, vol. 31, no. 4, Summer, 2015.

15. Andrew Brown, "When Evolutionists Attack," *The Guardian* . Mar. 6, 2006. http://www.theguardian.com/world/2006/mar/06/religion.uk.

16. Elliot Negin, "The 'Alar Scare' was Real," *Columbia Journalism Review* , 35, no. 3, (1996): 13 –15

17. Mary Lois Gibbs, *Love Canal: The Story Continues* (Washington: Island Press, 2011) and Adeline Gordon Levine, *Love Canal: Science, Politics, People* (Lanham, MD: Rowman & Littlefield, 1982).

18. Steven Epstein, *Impure Science: AIDS, Activism, and the Politics of Knowledge* (Berkeley: University of California Press, 1998).

19. National Academy of Science, *Women's Health Research: Problems, Progress, and Pitfalls*. (Washington, DC: National Academy Press, 2010).

20. Ned Block and Gerald Dworkin, *The IQ Controversy* (New York: Pantheon, 1976).

21. Mark Snyderman and Rothman Stanley, *The IQ Controversy: the Media and Public Policy* (Transaction Publishers, 1988).
22. Nicholas Rescher, *Cognitive Economy: Economic Perspectives in the Theory of Knowledge* (Pittsburgh: University of Pittsburgh Press, 1989).
23. Charles S. Peirce, "Note on the Theory of the Economy of Research," in *Collected Papers*, Vol. 7: 76–83. ed. Arthur W. Burks, 76–83 (Cambridge, MA: Harvard University Press, 1958).
24. William Kingdon Clifford, "The Ethics of Belief," *Contemporary Review*, 29 (December 1876–May 1877): 288–309.
25. Steve Fuller, "A Sense of Epistemic Agency Fit for Social Epistemology," chapter 2, section 4.
26. Steve Fuller, "The Dawn of Critical Neuroscience," *History of the Human Sciences* 07/2013; 26(3):107–15.
27. Jose Manuel Delgado, *Physical Control of the Mind: Toward a Psychocivilized Society* (New York: Harper, 1969).
28. Wilder Penfield, *Speech and Brain Mechanisms*, Princeton, Princeton University Press, 1959 and *The Mystery of the Mind* (Princeton, Princeton University Press, 1975).
29. Stanley Milgram, *Obedience to Authority: An Experimental View* (New York: Harper & Row, 1974) and Philip Zimbardo, *The Lucifer Effect: How Good People Turn Evil* (New York: Random House, 2008).
30. Arthur Caplan, ed. *When Medicine Went Wrong: Bioethics and the Holocaust* (Totowa, New Jersey, Humana Press, 1992) and Robert L. Berger, "Nazi Science: The Dachau Hypothermia Experiments," New England Journal of Medicine, 322, 20 (1990): 1435–40.
31. Stephen Downes, "Review of Science by Steve Fuller." *Philosophy of the Social Sciences* 30, 1 (2000): 140 –45.
32. Amos Tversky and Daniel Kahneman, "Judgment under uncertainty: Heuristics and biases," *Science* 185, 1124–31, 1974 and Amos Tversky and Daniel Kahneman, "Extensional versus intuitive reasoning: The conjunction fallacy in probabilistic reasoning," *Psychological Review* 90 (1983): 293–415.
33. Francis Remedios, Legitimizing Scientific Knowledge. Lanham, MD. Lexington Books, 2003:116.
34. Jay Labinger, *Up from Generality: How Inorganic Chemistry Finally Became a Respectable Field* (New York: Springer Verlag, 2013).
35. Pnina Abir-Am, "The Discourse of Physical Power and Biological Knowledge in the 1930s: A Reappraisal of the Rockefeller Foundation's 'Policy' in Molecular Biology," *Social Studies of Science* . 12, 3 (1983): 241 –82
36. Steve Fuller, "Response to Japanese Social Epistemologists: Some Ways Forward for the 21st Century," Social Epistemology 13, 3/4 (1999): 273.
37. Steve Fuller interview by Remedios, Buenos Aires, August 21, 2014.
38. Karl Marx, *The Eighteenth Brumaire of Louis Bonaparte* (New York: International Publishers, 1963).
39. Charles Webster, *The Great Instauration: Science, Medicine, and Social Reform 1626–1660* (London: Duckworth, 1975).

BIBLIOGRAPHY

Abir-Am, Pnina. "The Discourse of Physical Power and Biological Knowledge in the 1930s: A Reappraisal of the Rockefeller Foundation's 'Policy' in Molecular Biology." *Social Studies of Science* . 12, 3 (1983): 241 –82.
Anon. "You Can't Follow the Science Wars Without A Map." *The Economist*, Dec . 4, 1997. http://www.economist.com/node/109188.
Barnett, George, and Jack Otis. *Corporate Society and Education: The Philosophy of Elijah Jordan* . Ann Arbor: University of Michigan Press. 1961.

Barron, Colin. "A Strong Distinction between Humans and Non-Humans is no Longer Required
for Research Purposes: A Debate Between Bruno Latour and Steve Fuller." *History of the Human Sciences* 16 no. 2 (2003): 77–99.
Bell, Robert. *Impure Science: Fraud, Compromise, and Political Influence in Scientific Research*. New York: Wiley. 1992.
Berger, Robert L. "Nazi Science: The Dachau Hypothermia Experiments," New England Journal of Medicine, vol. 322, no. 20 (1990): 1435–40.
Block, Ned, and Gerald Dworkin. *The IQ Controversy*, New York: Pantheon. 1976.
Brown, Andrew. "When Evolutionists Attack." *The Guardian*. Mar. 6, 2006. http://www.theguardian.com/world/2006/mar/06/religion.uk.
Callon, Michel. "Some Elements of a Sociology of Translation: Domestication of the Scallops and the Fishermen of St Brieuc Bay." In *Power, Action and Belief: A New Sociology of Knowledge*:196–233. Edited by John Law. London: Routledge & Kegan Paul, 1986.
Caplan, Arthur, ed. *When Medicine Went Wrong: Bioethics and the Holocaust*. Totowa, New Jersey, Humana Press, 1992
Clifford, William Kingdon. "The Ethics of Belief," *Contemporary Review*, 29 (December 1876–May 1877) 288–309.
Delgado, Jose Manuel. *Physical Control of the Mind: Toward a Psychocivilized Society* . New York: Harper, 1969.
Downes, Stephen. "Review of *Science* by Steve Fuller." *Philosophy of the Social Sciences* 30, no. 1 (2000): 140 –45.
———. "Agents and Norms in a New Economics of Science." *Philosophy of the Social Sciences* 31 (2001): 224 –38.
Elgin, Caltherine. "Epistemic Agency." *Theory and Research in Education*. 11(2) (2013):135–52.
———. *Considered Judgment*. Princeton: Princeton University Press. 1999.
Elster, John. *Logic and Society*. Chicester, UK: John Wiley & Sons, 1979.
Epstein, Steven. *Impure Science: AIDS, Activism, and the Politics of Knowledge*. Berkeley: University of California Press, 1998
Fuller, Steve. *Social Epistemology*. Bloomington: Indiana University Press, 1988.
———. *Science*. Buckingham: Open University Press, 1997.
———. "Response to Japanese Social Epistemologists: Some Ways Forward for the 21st Century." Social Epistemology 13, 3/4 (1999): 273–302.
———. *Humanity 2.0. What it means to be Human Past, Present and Future*. New York and Basingstoke: Palgrave Macmillan, 2011.
———. "Social Epistemology: A Quarter Century Itinerary." *Social Epistemology* 26, 3–4 (2012): 267–83.
———. "The Dawn of Critical Neuroscience." *History of the Human Sciences* 07/2013; 26(3):107–15.
———. *Knowledge: The Philosophical Quest in History*. New York: Routledge, 2014a.
———. "Social Epistemology: The Future of an Unfulfilled Promise." *Social Epistemology Review and Reply Collective* 3, 7 (2014b): 29–37.
Gibbs, Mary Lois. *Love Canal: The Story Continues* . Washington: Island Press, 2011.
Goldman, Alvin. "Systems-Oriented Social Epistemology." *Oxford Studies in Epistemology*. Vol. 3: 189 –214. Edited by T. Gendler and J. Hawthorne. Oxford: Oxford University Press, 2010.
Gormley, Melinda, and Fellet, Mellisae. "The Pauling-Teller Debate: A Tangle of Expertise and Values. *Issues in Science and Technology* , vol. 31, no. 4, Summer, 2015.
Jordan, Elijah. *The Forms of Individuality*. Indianapolis: Progress Publishing Co., 1927.
Krimsky, Sheldon. *Science in the Private Interest: Has Corrupted Biomedical Research?* Rowman & Littlefield, 2004.
Labinger, Jay. *Up from Generality: How Inorganic Chemistry Finally Became a Respectable Field* , New York: Springer Verlag, 2013.

Lakatos, Imre. "Falsification and the Methodology of Scientific Research Programmes." *Criticism and the Growth of Knowledge*, 91–196, edited by Imre Lakatos and Alan Musgrave, Cambridge: Cambridge University Press, 1970.

Latou, Bruno. *Science in Action, How to Follow Scientists and Engineers through Society.* Cambridge, MA: Harvard University Press. 1987.

Levine, Adeline Gordon. *Love Canal: Science, Politics, People*, Rowman & Littlefield, 1982.

Manuel, Frank E. *The Prophets of Paris*. New York: Harper & Row, 1965.

Marx, Karl. *The Eighteenth Brumaire of Louis Bonaparte*. New York: International Publishers, 1963.

Milgram, Stanley. *Obedience to Authority: An Experimental View*, New York: Harper & Row, 1974.

National Academy of Science. *Women's Health Research: Problems, Progress, and Pitfalls.* Washington, DC: National Academy Press, 2010.

Negin, Elliot. "The 'Alar Scare' was Real." *Columbia Journalism Review*, 35, no. 3 (1996): 13–15.

Negley, Glenn. *The Organization of Knowledge*, New York: Prentice Hall, 1942.

Peirce, Charles S. "Note on the Theory of the Economy of Research." In *Collected Papers*, Vol. 7: 76–83. Edited by Arthur W. Burks, 76–83. Cambridge, MA: Harvard University Press, 1958.

Penfield, Wilder, *Speech and Brain Mechanisms*. Princeton, Princeton University Press, 1959.

———. *The Mystery of the Mind*, Princeton, Princeton University Press, 1975.

Remedios, Francis. *Legitimizing Scientific Knowledge*. Lanham, MD. Lexington Books, 2003.

———. "Review of *Knowledge: The Philosophical Quest in History.*" *Metascience*, forthcoming.

———. "Knowing Humanity in the Social World: A Social Epistemology Collective Vision?" In *The Future of Social Epistemology*:21-27. Edited by James Collier. London: Rowman & Littlefield International, 2015.

Rescher, Nicholas. *Cognitive Economy: Economic Perspectives in the Theory of Knowledge.* Pittsburgh: University of Pittsburgh Press, 1989.

Roth, Paul. "The Bureaucratic Turn: Weber contra Hempel in Fuller's Social Epistemology." *Inquiry* 34 (1991):365–76.

Snyderman, Mark, and Stanley, Rothman. *The IQ Controversy: the Media and Public Policy*. Transaction Publishers, 1988.

Taubes, Gary, "Postol versus the Pentagon," *Technology Review*. April, 2002.http://www.technologyreview.com/featuredstory/401412/postol-vs-the-pentagon/.

Tversky, Amos, and Kahneman, Daniel. "Judgment under uncertainty: Heuristics and Biases." *Science* 185, 1124–31, 1974.

———. "Extensional versus intuitive reasoning: The conjunction fallacy in probabilistic reasoning," *Psychological Review* 90, 293–415, 1983.

Uebel, Thomas. "Review of Legitimizing Scientific Knowledge: An Introduction to Steve Fuller's Social Epistemology." In *Notre Dame Philosophical Reviews* 2005. http://ndpr.nd.edu/news/24412-legitimizing-scientific-knowledge-an-introduction   to-steve-fuller-s-social-epistemology/.

Webster, Charles, *The Great Instauration: Science, Medicine, and Social Reform 1626–1660.* London: Duckworth, 1975.

Wennemann, Daryl. *Posthuman Personhood*. Lanham, MD: University Press of America, 2013.

Zimbardo, Philip. *The Lucifer Effect: How Good People Turn Evil*. New York: Random House, 2008.

FIVE
Agency and Disagreement

Paul Faulkner

In his opening chapter "A Proposed Research Program for Social Epistemology," Sanford Goldberg embraces the distinction between *believing what a speaker says* and *believing a speaker*.[1] In his terms this is a distinction between treating a speaker's testimony as a piece of evidence and accepting the speaker's "word for something." In the latter case, we take speakers "not merely as providing potential evidence, but also, and more centrally, as manifesting *the very results of their own epistemic sensibility*."[2] Social epistemology, Goldberg proposes, should then abandon the individualistic framework implicit in the evidential view and focus on the connection forged by believing the speaker; where this is something, Goldberg notes, that he will not argue for but "would like to suggest how social epistemology looks from the vantage point of those who take this [latter] picture seriously."[3] The project is then to account for how taking a speaker's word for something 'socially distributes' the acquisition of knowledge. I'd like to follow Goldberg in taking this distinction as my unargued starting point.[4] But from this starting point my concern is to pursue another of Goldberg's suggestions: the idea that epistemic agency then enters in "the various roles that epistemic subjects play in acquiring, storing, processing, transmitting or assessing information."[5] I think that this is correct, and I want to consider the specific manifestation of agency that is choosing whether to take a speaker's word for something or to treat the speaker's testimony as just another piece of evidence.

The focus will be on expert testimony, and specifically on occasions where the expert's opinion is something of importance to an audience who is thereby at pains to assess the credentials of the expert and pays careful attention to what is said. Such occasions are common; they are confronted by most, for instance, whenever a tradesman is engaged or the services of a lawyer, doctor, or financial advisor are employed. I'll proceed with some examples and details below. But for now it can be noted that a feature of these cases is that one will often have good reasons to believe what the expert says; it might be, say, that a tradesman has been recommended on their track record and their assessment of the job strikes one as careful and about right. And yet equally, at a certain point, it seems as though one can choose to put one's faith in the tradesman as expert and believe what they say *because one believes them*. In these cases there are two epistemically permissible routes to belief. The audience who has carefully assessed an expert's testimony can treat it as a bit of evidence and believe what is said on the basis of the reasons supplied by their careful assessment. Or the audience can believe the expert, and take their word for things. Both routes to testimonial uptake are epistemically permissible and, as such, agency is implicated in *how* a testimonial belief is formed. Thus, the first ambition of this chapter is to show that, on those occasions where there is a live distinction between believing a speaker and merely believing what the speaker says, our testimonial uptake manifests our agency, and specifically *epistemic* agency.

The second ambition of this chapter is to show how this sheds light on the epistemology of disagreement. It does so because how an audience chooses to respond to the testimony of an expert determines the audience's epistemic position were they to subsequently encounter another expert with a different opinion. Roughly, I will argue that were the audience to have taken the first expert's word for it, suspension of belief would be demanded on encountering the second expert. But if the audience treated the first expert's testimony as no more than a piece of evidence, then there would be some room for the audience to reject the second expert's opinion and hold fast to their belief. However, this encounter with the second expert simply generates a third person variant of the first person case of disagreement that has perplexed the literature — that is, a case where one disagrees with an epistemic peer.[6] Thus the result of considering these testimonial cases is some account of what is epistemically needed to hold fast in the face of disagreement.

This chapter is then structured as follows. In the next section, I develop the distinction between believing a speaker and believing what a speaker says, show how it entails a significant place for agency within testimonial uptake, and consider whether this distinction can be accommodated within Goldberg's epistemological framework. In the following section I outline the implications of this agency for the epistemology of disagreement. Then in the final section I explain how this agency is decidedly epistemic in character.

1. TESTIMONIAL UPTAKE AND AUDIENCE AGENCY

As audiences we can treat testimony as evidence or take speakers' words for things. Before adding some flesh to this distinction, it might help to have a specific case in mind, so consider some engine trouble.

> *Case 1, the driver.* Her car keeps stalling and she doesn't know why, possessing no car-mechanical knowledge. Thankfully this is less troublesome than it could be because there is a mechanic she always goes to and, as far as she can tell, he has always done a good job. The mechanic tells her that her carburetor needs to be rebuilt.[7]

> *Case 2, the middle-aged runner.* At a recent local race he felt his heart flutter, and this was subsequently diagnosed as an episode of atrial fibrillation. After some enquiry and research, he discovered a number of other runners with the condition who all recommended visiting a certain cardiologist. This cardiologist's verdict was that it was safe to continue running but racing needed to stop.

In both these cases, the audience relates to the speaker as layperson to expert. That is, while the audience might have some idea of what a carburetor is or what atrial fibrillation amounts to, in neither case will the audience know what evidence supports the expert verdict, nor what could establish the truth of this verdict.

Nevertheless, this epistemic limitation does not mean that these audiences cannot evaluate these bits of expert testimony.[8] Indeed, in both these cases the audience has information that allows them to evaluate whether the testimony is true, or likely to be so. In case one, the information is the driver's observation of the mechanic's track record: given this, she can reasonably judge it to be true that her carburetor needs rebuilding. In case two, the information is the testimony the runner has received as to the cardiologist's track record: given this, he can equally judge it to be safe to run if not race. While the aim of evaluating a piece of testimony is to form some judgment as to its truth, in these cases where the audience doesn't have a prior belief as to the truth of what is said, this evaluation proceeds by way of a judgment as to whether the given bit of testimony is *evidence for its truth*.

Having then evaluated a piece of testimony, an audience is then in the epistemic position where the audience can believe what is said on the basis of this evaluation. Were the audience to do this, and thereby treat a bit of testimony merely as piece of evidence, the audience would treat it like any other natural phenomena. A speaker's statement that her daughter has measles would, on this approach for instance, be comparable to the observation of her daughter's measles rash: both would be simply evidence of the child's measles.[9] To respond to testimony in this way would be to then approach it *objectively* not as a participant in the conversational exchange but as an observer of it. On this approach, a speaker's intentions can be irrelevant; if I have established a strong correlation between one speaker lying and being asked whether some thing is so, and between another speaker telling the truth and being asked whether the same thing is so, then the testimony of each is as good as the testimony of the other even though each will have a quite different intentional explanation. As such, this approach allows an audience to maintain a degree of *epistemic autonomy* since what gives the audience their reason for belief is *their background of belief* that supports the judgment that the bit of testimony is a piece of evidence.[10]

Things don't need to proceed this way. The driver might believe that her carburetor needs rebuilding because she believes the mechanic. According to Richard Moran believing a person involves believing what that person tells one merely because one recognizes their telling to be a certain kind of intentional act.[11] To explain the kind of intentional act it is, Moran appeals to Grice's distinction between 'telling' and 'letting know' and the difference between photographs and sketches. Suppose that a mother sees her daughter has a rash and recognizes that this is a measles rash. The observation of the rash *lets the mother know* her daughter has measles, and a photograph of the rash, shown to her friend who is a doctor, would let the doctor know the same thing. However, if the mother didn't have a photograph to hand and had to draw the rash for her friend, then the sketch gives this friend a reason to believe the daughter has measles only insofar as she worked out that the mother's intention was to depict her daughter's rash; that is, only insofar as the friend recognizes the sketch as a certain kind of intentional act. The same is true, Moran argues, of *tellings*. Thus, in believing the mechanic when he tells her that the carburetor needs rebuilding, the driver's reason for belief comes from her recognizing the mechanic's testimony as a certain intentional act: an *assurance* that what is said is true. To believe a speaker is to be assured and so believe what is said *on the speaker's authority*; it is to *defer* to the speaker as an authority.[12] Moreover, in telling the driver what he does, the mechanic then presents himself as an authority and expects the driver to believe that her carburetor needs rebuilding *simply because this is what he tells her*.

What I would add to this account is that our fulfilling these speaker expectations involves *trusting* the speaker.[13] In this case, although the driver has evaluated the mechanic's credentials, her response to his testimony can still be trusting in that she can recognize that ultimately she doesn't know how to verify or falsify the mechanic's judgment. Trust then involves taking the mechanic's telling at face value as the honest diagnosis it purports to be. And in then deferring to the mechanic and believing what he tells her on his authority, or for the reason that he intends, she thereby engages with the conversational exchange *as a participant*; which is to say that she believes that her carburetor needs rebuilding for the reason that the mechanic intends she believe this: his telling it.[14]

One can believe what a speaker says either on the basis of treating the speaker's testimony as a piece of evidence or on the basis of believing the speaker.[15] The foregoing account of this distinction is sketchy at best, but, as stated, I am going to largely take this distinction for granted.[16] Nevertheless, the foregoing should have established that in treating a bit of testimony as evidence one takes an objective stance on a conversation and maintains a degree of epistemic autonomy; that is, for instance, one treats a bit of testimony as one might treat a measles rash, or photograph of this rash: as evidence that something is so. By contrast, in believing a speaker, one takes a participant stance in a conversation and defers to the speaker's authority. Thus, and for instance, the middle aged runner could simply add the cardiologist's statement to the stock of information he has gathered in researching what is safe for him to do with his condition; or he could trust the cardiologist. Whether he defers or keeps his own counsel is then a choice he makes. It is a manifestation of his agency. Moreover, in this case at least, where there has been some positive evaluation of the speaker's testimony, the choice is between two routes to belief that are *equally epistemically permissible*. That is to say, both ways of forming belief will result in a belief that is rationally supported or justified. As such, the agency exercised is, one might say, *epistemic agency*.

That epistemic agency can be exercised in this context is not a fact that can be captured by all theories. In particular, it is not clear it can be captured by reliabilist theories—such as Goldberg's. A question for any reliabilist theory is the identity of the process whose reliability is to be assessed.[17] On this matter Goldberg has changed his view from taking the process to be an individual audience's receipt of testimony to taking it to be the "extended process" that includes both the production and the receipt of testimony.[18] On either account, insofar as justification is solely a function of the reliability of the process of belief formation, whether it is conceived in extended terms or not, if there are two unique processes that would result in the same belief, as there are in case 1 and case 2 above, *the subject should employ the more reliable process*. Thus there will be a reliability associated, in each of these cases, with taking the expert verdict as evidence and taking in on trust. These facts about reliability will then determine how the driver and runner in these cases *ought* to form belief. So while each might have the capacity to choose their process of belief formation, neither is at epistemic liberty to do so. This is not a case of *epistemic agency* as such.

The obvious worry, if this is true, is that this criticism of reliabilism can be generalized to be a criticism of any theory of epistemic justification. Thus, and for instance, suppose one endorses the justificationist view that being justified consists in a certain standing in the space of reasons. These two routes to belief surely determine different degrees of rational support. But then doesn't it equally follow that an audience ought to form their belief in the way that ensures it receives the greatest rational support? And if so, there is again choice but not epistemic agency because what one ought to choose is clear from the epistemic perspective: that route to belief which achieves the best standing in the space of reasons. This statement of the worry, I will argue, misses the point that one can have different kinds of epistemic reason. There are reasons for the runner, say, to strive to maintain epistemic autonomy, and there are reasons for the runner to trust, and both these reasons can be epistemic. To substantiate these claims then involves an excursion into the epistemology of disagreement.

2. THE EPISTEMOLOGY OF DISAGREEMENT

The epistemology of disagreement is concerned with cases like the following.

> Case 3. *The restaurant case.* Five of us go out to dinner and agree to split the bill evenly adding a twenty percent tip. After doing the mental arithmetic, I conclude that we each owe $43, you conclude it to be $45.[19]

Here the epistemic issues start with the presumption that you and I are in comparable epistemic positions when it comes to the question of how much we each owe. It is this presumption that makes our disagreement something of a puzzle. There would be no such puzzle if our epistemic positions were not comparable—for instance, if I used a calculator and you were six years old—but they are comparable: we are both sober (enough) and the arithmetic is simple. In short, *we are in comparably equivalent epistemic positions with respect to the disputed question*, which in this case is the money we each owe. Where this is true, let me say that we are *epistemic peers* with respect to this question.[20] So disagreement raises an epistemic question when it is with someone whom one *knows* to be one's epistemic peer. However, it would equally raise an issue were one merely to *believe*, even if incorrectly, that another was one's epistemic peer. For instance, if you were much more intoxicated than you seem, and had lost the ability to do mental arithmetic, then all the while I am unaware of this, and so believe you to be my peer, your reaching a different sum, would still raise the issue of how I should respond.[21]

In the restaurant case, given that I think of you as my peer, and not an expert, I won't simply assume you're right. So my options seem to be two. Either I assume that we are both as likely to be wrong as right, and I suspend judgment pending a recalculation. Or I assume, pending a recalculation, that you must have made a mistake. The *equal weight* or *conformist* position is that the former assumption is the epistemically correct one to make. One should be moved to suspend judgment, Elga says, because "one should give the same weight to one's own assessments as one gives to the assessments of those one counts as one's epistemic peers."[22] The idea behind the alternative *non-conformist* view is that if I have *in fact calculated correctly*, this fact should make some epistemic difference. Thus Kelly argues that it is the evidence that determines the correct epistemic response in cases of disagreement. In the restaurant case, your testimony is "higher-order evidence" in that it is "evidence to the effect that one is currently in circumstances in which one is more likely than usual to have made a mistake."[23] But it is the evidence, taken as a whole, which determines the right response. Thus, Kelly writes,

> what it is reasonable to believe depends on both the original, first-order evidence as well as on the higher-order evidence that is afforded by the fact that one's peers believe as they do. For this reason, it seems appropriate to call the view on offer the Total Evidence View.[24]

So if I have "in fact correctly evaluated" the situation, my belief that we owe $43 enjoys a certain epistemic standing.[25] This kind of epistemic standing can be sufficient, though maybe not in this case, to render it "reasonable" to hold steadfast to my belief in the face of disagreement.

A problem for this non-conformist view is that it can be unreasonable to be steadfast in belief even when one enjoys such a first rate epistemic standing as *knowing* what one believes to be the case. Imagining a third party instance of disagreement shows this. (In a standard disagreement situation, such as case 3, one party A believes that p, and then confronts testimony to not-p from a second party B, who A believes to be an epistemic peer; a third party case is one where a third subject, who believes A and B to be epistemic peers, first encounters A's testimony to p, and then confronts B to not-p.[26]) For instance, imagine the following continuation of case 1.

> Case 1*. *The driver case continued.* Having been told her carburetor needs rebuilding (at time t_1), but wary of the cost, the driver decides to get a second opinion and takes her car to another mechanic who has been highly recommended. He says (at time t_2) that the spark plugs need changing, and nothing more need be done.

Suppose two further things about this case. First, that the driver's response to the first mechanic's testimony is to trust the mechanic and believe that her carburetor needs to be rebuilt on his authority. Second, that the mechanic knows what he tells her. Given these suppositions, the driver knows, at t_1, that her carburetor needs rebuilding (call this proposition 'p'). She knows that p in the ordinary way that we can know things on the basis of expert testimony: by trusting an expert who is in authoritative. However, this epistemic standing—knowing that p—is not sufficient to make it "reasonable" for the driver to ignore the second mechanic's testimony that not-p. The reason that it is not sufficient, and that it would be dogmatic to ignore this second piece of testimony, is that it is *merely a historical accident* that the driver believes that p rather than not-p due to the fact of taking her car to the first mechanic first.[27] That is, had the driver reversed the order of her garage visits, she would have believed that spark plugs were the issue at t_1. So two things are true of the continued case 1*: the driver knows that p at t_1; but the epistemically correct response to the second mechanic's testimony at t_2 is to suspend this belief.

However, one should not conclude that the equal weight or conformist view is correct because while *knowing* that p is not sufficient to reasonably persist in believing that p in the continued driver case (case 1*) in other situations this epistemic standing does seem to render steadfastness in belief reasonable. For instance, imagine a continuation of case 3:

> Case 3*. *The restaurant case continued.* After several recalculations I still come to the conclusion that we all owe $43. So I get a pencil and paper and lay out my sums, I cannot see any error and yet you still claim it to be $45.

Or consider—and call it case 4, *the university quad*—a case described by Richard Feldman:

> Suppose you and I are standing by the window looking out on the quad. We think we have comparable vision and we know each other to be honest. I seem to see what looks to me like the Dean standing out in the middle of the quad. (Assume that this is not something odd. He's out there a fair amount.) I believe that the Dean is standing on the quad. Meanwhile, you seem to see nothing of the kind there. You think that no one, and thus not the Dean, is standing in the middle of the quad. We disagree.[28]

Feldman uses this case to argue for the equal weight or conformist view. What I should do in this situation is "suspend judgment." However, this conclusion is both descriptively and normatively problematic *if* it is plausibly supposed that the Dean is in fact standing there, and the scene is perceptually vivid; if it is supposed, as Ernest Sosa elaborates, that "the quad is small and sunlit, and the dean sports a bright blue jacket just fifty feet away."[29] Given this supposition, it would take more than your testimony to make me suspend my belief that the Dean is in the quad because there is something compelling about my vivid perception of this fact. Put in the strongest terms, my believing this is, as Catherine Elgin remarks, not something I "could refrain from doing at will."[30] But more than this, in this situation my perceptually given knowledge that the Dean is in the quad does seem to make it reasonable for me to persist in believing this.

The issue then is why knowing that p can make it reasonable to reject testimony to not-p on some occasions (like 3*) and not others (like 1*).

Jennifer Lackey, who considers two cases very much like these last two, proposes on their basis the following principle.

> No Doxastic Revision Required: In a case of ordinary disagreement between A and B, if A's belief that p enjoys a very high degree of justified confidence, then A is permitted to rationally retain her same degree of belief if and only if A has a relevant symmetry breaker.[31]

Justification, for Lackey, entails reliability.[32] So "justified" in this quote flags a success state similar to Kelly's 'correct evaluation of the situation.' "Confidence" is then supported by and entails the possession of "person information" that acts as the relevant symmetry breaker. For example, with respect to case 4, there are two explanations of your utterance denying the Dean's presence: the explanation that starts from the Dean not being there and proceeds to hypothesize an unusual error on my part; and the explanation that starts from the Dean in fact being there and proceeds to hypothesize an unusual error on your part. What allows me to favor the latter and infer your error as the best explanation of your utterance is the personal information I possess "about the normal functioning of [my] own cognitive faculties."[33] For instance, I know I am sober, not depressed, wearing my glasses and so on.

The problem with this account is that it would seem simple to beef up this case so that whatever personal information I possess about myself, I equally possess about you. Couldn't I know you to be similarly sober, non-depressed, wearing your glasses and so on? That is, any piece of personal information I might possess about the functioning of my faculties I might equally possess, from the third person perspective, about the functioning of your faculties.

Nevertheless, I think Lackey is right to say that *confidence*, in conjunction with some objectively good epistemic standing, matters. In these cases where the steadfast response seems reasonable, it is not merely that one knows but it is further that one is confident that one knows. It is just that reference to personal information doesn't provide a good explanation of the ground and nature of confidence. The question is then what does?

Again I think that a third person case can help here. So consider the following continuation of case 2.

> Case 2*. *The middle aged runner continued*. Having been told that it is safe to run but not race, at time t_1, the middle-aged runner has a follow up appointment and sees a different consultant. At time t_2, he is told that there is no safe level of risk, and it is best not to run at all.

Suppose two further things about this case. First, the runner's response to the first consultant is not to believe what he is told on the basis of trust, but to believe it on the basis of a careful and correct evaluation of the consultant's testimony as a piece of evidence. Second, the first cardiologist is correct in her assessment. Given these suppositions the runner knows, at t_1, that it is safe to run but not race (call this proposition 'p'). He knows that p in the ordinary way that we can know something on the basis of the evidence, and can get to know things on the evidence of testimony.

This epistemic standing—knowing on this articulated evidential basis—offers some grounds for the reasonable rejection of the second consultant's testimony to not-p encountered at t_2. First, while it may be a historical accident that the runner encountered the first cardiologist first, this does not entail, as it did in the trust case, that it would be dogmatic to persist in belief. This entailment is broken by the fact that different histories of evidential accumulation "make a causal difference to which body of total evidence one ultimately ends up with."[34] So encountering the first cardiologist first makes an evidential difference. It does so, Kelly argues, because one is prone to scrutinize counter-evidence, or evidence that tells against existing belief, through looking for logical fallacies, methodological improprieties and, crucially, imagining alternative explanations of the facts under which they lose their counter-evidential value. This habit, Kelly argues, is justified because the requirement that supporting evidence be equally scrutinized would be "extremely demanding."[35] (For instance, having established that all ravens are black, it would be too demanding to require that one continue to check every black raven; it is non-black ravens that need to scrutinised.)

Moreover, second, the fact that the runner himself gave a knowledge supporting argument for that p entails, other things being equal, that it is reasonable for the runner to believe that p; and *ipso facto* it makes it reasonable for him to start reasoning from the presumption that not-p is false. This is not to suggest that it is reasonable to reject the second cardiologist's testimony out of hand—to do this would be dogmatic. Rather, it is to claim that it is reasonable for the runner to *first presume* the second cardiologist has got things wrong and try to imagine alternative explanations of her testimony that would support this presumption. In doing this, the runner tries to imagine a plausible symmetry breaker. Were this exercise unsuccessful, and assuming the runner took both cardiologists to be peers it probably would be so, the only reasonable response would be to suspend belief. But there is a starting asymmetry: the fact that the runner has already articulated a knowledge supporting argument for p, makes it reasonable to start with an attempt to explain why the second cardiologist's testimony is in fact misleading.

Sometimes it is reasonable to hold fast to one's belief in the face of disagreement, and sometimes it is not. This difference is illustrated in a comparison of the two restaurant cases. In the initial situation (case 3), suspension of belief is the reasonable response, but in the continued case (case 3*), it seems reasonable to persist with one belief. Comparison of two third party cases of disagreement—cases 1* and 2*—then suggests an explanation of when and why persistence is reasonable. In both of these third party cases the subject approaches the contrary testimony, given at t_2, knowing that p on the basis of testimony received at t_1, but the difference in the basis of testimonial knowledge seen across the two cases generates a difference in the reasonableness of continuing to believe that p. To elaborate this a few terms of art are needed. Let 'warrant' be a generic term referring to any species or degree of epistemic support. The epistemic character of warrant is then determined along dimensions of *strength* and *transparency*. Strength is simply a matter of the degree of epistemic support warrant gives. While transparency is a matter of a believer's access to the facts that determine warrant and awareness of the support these facts give. So if a subject S believes that p and S's belief that p enjoys warrant W on the basis of facts F, then *the strength of W is the probability that p is true given F*; and *the transparency of W is the degree to which S is aware of F and aware that F determines W*. Insofar as the truth of the proposition believed is determined by objective fact, strength could be conceived in terms of objective probability. As such strength is an externalist notion, which correlates with reliability. While transparency in being defined in terms of 'access to' and 'awareness of' the warranting facts and that these facts warrant, is an internalist notion. The simple cases that tell against the equal weight or conformist view—such as the continued restaurant case 3*—involve situations where the subject's warrant is both strong (sufficient for near certainty) *and* transparent (the subject is aware of what grounds their knowledge and that it does so). What the two third party disagreement cases—cases 1* and 2*—then illustrate is that these two dimensions of epistemic warrant can be separated.

In believing the first mechanic, in cases 1 and 1*, and so getting to know that her carburetor needs to be rebuilt, the driver's warrant for belief is strong: insofar as she believes this on his authority, her belief inherits his warrant. But this warrant is entirely non-transparent to her. That is, she has no access to, or comprehension of, the facts that determine the warrant of her testimonial belief. On might say, following Hardwig, that knowledge acquired on trust is *blind*.[36] By contrast, in believing what the first cardiologist says, in cases 2 and 2*, and so getting to know that it is safe to run, the runner's warrant for this belief is not particularly strong: it is merely whatever grounds his positive evaluation of the cardiologists testimony. In particular, in treating the cardiologist's testimony as a piece of evidence, the runner does not inherit the much stronger warrant that the cardiologist has for what she tells him. The runner does not inherit this warrant because he expressly does not believe this fact on her authority, but believes it on the basis of judging her to be authoritative. But the benefit of maintaining his epistemic autonomy—of having his warrant for belief grounded on other things he believes—is that this warrant is entirely transparent.[37] Since it is only in case 2* that there is a possibility of reasonable persistence in belief, what is necessary for this possibility is that one's warrant for the persisting belief be transparent—as it is in case 2*. That is, it is the transparency of warrant that determines matters and not its strength.

Suppose, then, that cases like 3* and 4, the continued restaurant and university quad cases, describe situations where the believer has, as Lackey suggests, a *justified confidence* in their belief. What underwrites this confidence, I want to suggest, is the fact that the warrant possessed in these cases is possessed transparently. And it is the holding of this fact that is necessary for any reasonable persistence of belief in the face of disagreement.[38]

3. EPISTEMIC AGENCY

When a speaker tells one that p, it can be that an audience has a choice. In those cases where this choice is live—such as the driver and runner cases 1 and 2—an audience can believe that p, where this is what the relevant speaker says, because they *believe the speaker*; or they can *take the speaker's testimony as a piece of evidence* that indicates that p. The uptake of such testimony thereby manifests an audience's agency. And this agency can be regarded as epistemic insofar as both choices are epistemically permitted. This is true of these two key cases because both routes to believing that p, in these situations, yield a warranted belief that p. That the driver (in case 1) and the runner (in case 2) have this *epistemic* liberty is then a fact that cannot be captured by reliabilism. This is because in each case there will be some fact of the matter as to what route to believing that p is the most reliable. Since this route yields the belief with the greater warrant, it is this route to belief that is epistemically mandated. The same problem equally confronts any theory of warrant, which like reliabilism, acknowledges only one dimension of warrant: such theories must conclude that one route to belief, in these cases, is epistemically mandated.[39] However, the foregoing discussion of disagreement has hopefully then shown that there is an epistemic rationale for defending the claims that both routes to belief yield warranted. That is to say, there is an epistemic reason for preferring a *strongly warranted* belief, and an epistemic reason for preferring a *transparently warranted* belief.

Being epistemically autonomous is a matter of not depending on others for warrant. In being epistemically autonomous one is in a position to explain how it is that one knows what in fact one does know. This is because the warrant that determines this epistemic standing will be transparent. Having a transparently warranted belief then grounds the possibility of reasonable persistence of belief in the face of disagreement. It does so because it makes it reasonable for one to start one's reasoning about the disagreement from the presumption that the other party got things wrong—from the presumption that this counter-evidence is misleading. Given that epistemic autonomy delivers this entitlement, it is unsurprising that epistemologists have traditionally regarded epistemic autonomy as valuable and something that one should strive for. The motto of *The Royal Society of London*—established in 1600 as a center for the new science—was *On No Man's Word*. And in the same spirit Robert Boyle, the paradigmatic new experimental philosopher, claimed "The great Reverence men usually give to humane Authority is undeserved. . . . Humane testimony ought not to be a force against either Reason or Experience."[40]

However, in contemporary epistemology such a view is usually criticized for its rampant individualism—Goldberg, for instance, rejects this individualism out of hand as it "fundamentally mischaracterizes" our epistemic relation to others.[41] And this claim can be epistemically supported too. In a testimonial context, the bid to maintain one's epistemic autonomy is impoverishing in the sense that it results in our having beliefs whose warrant is weaker than it could be. For instance take case 2. The warrant for believing that it is safe to run that is possessed by the cardiologist is stronger than the warrant that is possessed for believing the same proposition on the basis of crude assessments of the cardiologist's reliability. It is stronger because only the cardiologist's expertise takes into account the facts that make it true that it is safe to run. So there is also normative pressure to trust. This pressure is partly social: there is a social norm dictating that one ought to tell the truth and believe others.[42] But it is also epistemic: one ought to base one's belief on the best possible available evidence, and this evidence is often that possessed by other people. It is evidence we only have access to when we believe on others' authority. This, I think and in conclusion, is then a feature of our epistemic lives: there are different and conflicting epistemic imperatives. 'Maximize the truths believed!' and 'Minimize the falsehoods believed!' are two. 'Be epistemically autonomous!' and 'Trust!' are two more. Our epistemic agency is then manifested in choosing which directive to follow.

NOTES

1. Thanks are owed to Rosanna Keefe, Mari Mikkola, David Owens, Patrick J. Reider, Bob Stern, and audiences at the University of Reading, University of Copenhagen, and Humboldt University, Berlin where previous versions of the paper were presented.
2. Goldberg, chapter one.
3. Goldberg, chapter one.
4. I argue at length for this distinction in (Faulkner 2011).
5. Goldberg, chapter one.
6. See, for instance, (Feldman and Warfield 2010).
7. Example from (Hardwig 1985), p. 335.
8. A point that has been successfully argued by (Goldman 2001). For a general account of how we evaluate the truth of testimony, see (Faulkner 2011), chapter 2.
9. Both, Grice would say, are treated as "natural" signs, (Grice 1957), p. 214.

10. A point emphasized by (McMyler 2011).
11. See (Moran 2006).
12. Another point emphasized by (McMyler 2011).
13. See (Faulkner 2011), ch.6.
14. Following Holton, one could say she takes the 'participant stance,' (Holton 1994), p. 66.
15. See also (Zagzebski 2012), pp.124–25 for the claim that testimony can be treated in these two ways.
16. To detail it properly would require articulating complete reductive and assurance accounts of testimony, again see (Faulkner 2011) chapters 2 and 6 respectively.
17. On Lackey's view, for example, it is the reliability of the testimony itself—the reliability of the speaker's words that matter. Since this would be constant whether, say, the runner treated the cardiologist's statement as evidence or trusted the cardiologist, there is no epistemic significance to this distinction.
18. For the old view see (Goldberg 2007), where this view is shared by (Graham 2000); for the new view see (Goldberg 2010).
19. David Christensen. "Epistemology of Disagreement: The Good News," *Philosophical Review* 116 (2) (2007): 193.
20. The requirement is not that our epistemic positions be identical, for this they never could be. It is just that, under some measure, they be equivalent. For similar but different definitions see (Elga 2011), p. 167, n.1; (Feldman 2011), p. 144; (Kelly 2005), pp. 174–75; (Lackey 2010), p. 303, and (Sosa 2010), p. 283.
21. Elga equally focuses on the situation where I "count" you as an epistemic peer (Elga 2011), p.163. And Lackey focuses on cases of "ordinary disagreement' where disputants "regard each other" as epistemic peers, rather than "idealized disagreement" where disputants are in fact peers (Lackey 2010), p. 304.
22. Adam Elga, "Reflection and Disagreement." In *Social Epistemology: Essential Readings*, ed. Alvin Goldman and Dennis Whitcomb, 158–82 (Oxford: Oxford University Press 2007), 164; see also (Christensen 2007) and (Feldman 2011).
23. Thomas Kelly, "Peer Disagreement and Higher-Order Evidence." In Disagreement, ed. R. Feldman and Ted Warfield (Oxford: Oxford University Press 2010), 139.
24. Ibid., 142.
25. Thomas Kelly, "The Epistemic Significance of Disagreement." In Oxford Studies in Epistemology, ed. Tamar Szabo Gendler and John Hawthorne (Oxford: Oxford University Press 2005), 180.
26. This third party case is of independent interest because it raises the question of which experts one should believe.
27. Compare (Kelly 2008), p. 616.
28. Richard Feldman, "Reasonable Religious Disagreements." In Social Epistemology: Essential Readings, ed. Alvin Goldman and Dennis Whitcomb (Oxford: Oxford University Press), 151.
29. Ernest Sosa, "The Epistemology of Disagreement." In Social Epistemology, edited by A. Haddock, A. Millar, and D. Pritchard (Oxford: Oxford University Press 2010), 293.
30. Catherine Z. Elgin, "Persistent Disagreement." In Disagreement, edited by R. Feldman and Ted Warfield (Oxford: Oxford University Press, 2010), 60. Of course, Elgin here follows Hume.
31. Jennifer Lackey. "A Justificationist's View of Disagreement's Epistemic Significance." In *Social Epistemology*, ed. A. Haddock, A. Millar, and D. Pritchard (Oxford: Oxford University Press, 2010), 319. Lackey's cases are 'elementary math,' p. 306, and 'perception' respectively, p. 308.
32. Ibid., 320.
33. Ibid., 309.
34. Kelly, "Peer Disagreement and Higher-Order Evidence," 628.
35. Ibid., 623.

36. J. Hardwig, "The Role of Trust in Knowledge." *The Journal of Philosophy* 88 (12) (1991): 699.

37. It is worth remarking that the way I am using 'transparent' is different from another established usage according to which it refers to the fact that any deliberation about whether to believe that p is deliberation about whether p is true, see (Shah 2006), p. 481.

38. Kelly is sensitive to, though does not develop, this point. He observes that "it is the fact that the status of one's performance is not perfectly transparent that opens the door for higher-order considerations to make a difference" (Kelly 2010), p. 140.

39. Indeed, this is the point of Goldman's veritism: to assess which, of various social routes to belief, maximizes veritistic value (or is the most reliable). See (Goldman 1999).

40. Quoted in (Shapin 1994), p. 202.

41. Goldberg, chapter one. Or see, for example (Coady 1992), pp. 13–14.

42. See (Faulkner 2010) and (Faulkner 2011), §7.3.

REFERENCES

Christensen, David. "Epistemology of Disagreement: The Good News." *Philosophical Review* 116 (2) (2007):187–217.

Coady, C.A.J. *Testimony: A Philosophical Study*. First ed. Oxford: Clarendon Press, 1992.

Elga, Adam. "Reflection and Disagreement." In *Social Epistemology: Essential Readings*, edited by Alvin Goldman and Dennis Whitcomb, 158–82. Oxford: Oxford University Press. Original edition, 2007.

Elgin, Catherine Z. "Persistent Disagreement." In *Disagreement*, edited by R. Feldman and Ted Warfield, 53–68. Oxford: Oxford University Press, 2010.

Faulkner, Paul. "Norms of Trust." In *Social Epistemology*, edited by D. Pritchard, A. Haddock, and A. Millar, 129–47. Oxford: Oxford University Press 2010.

Faulkner, Paul. *Knowledge on Trust*. Oxford: Oxford University Press, 2011.

Feldman, Richard. "Reasonable Religious Disagreements." In *Social Epistemology: Essential Readings*, edited by Alvin Goldman and Dennis Whitcomb, 137–57. Oxford: Oxford University Press, 2011.

Feldman, Richard, and Ted Warfield, eds. 2010. *Disagreement*. Oxford: OUP.

Goldberg, Sandford. "A Proposed Research Program for Social Epistemology." In *Social Epistemology and Epistemic Agency*, edited by P. Reider. Rowman & Littlefield, 2016.

Goldberg, Sanford. *Anti-Individualism: Mind and Language, Knowledge and Justification*. Cambridge: CUP, 2007.

Goldberg, Sanford. 2010. *Relying on Others*. Oxford: Oxford University Press.

Goldman, A. "Experts: Which Ones Should You Trust?" In *Pathways to Knowledge: Private and Public*, edited by Goldman. Oxford: Oxford University Press, 2001.

Goldman, Alvin. *Knowledge in a Social World*. Oxford: Clarendon Press, 1999.

Graham, Peter J. "The Reliability of Testimony." *Philosophy and Phenomenological Research* 61 (3):695–709, 2000.

Grice, P. "Meaning." In *Studies in the Way of Words*, edited by P. Grice, 213–23. Cambridge, MA: Harvard University Press, 1957.

Hardwig, J. "Epistemic Dependence." *The Journal of Philosophy* 82 (7) (1985):335–49.

Hardwig, J. "The Role of Trust in Knowledge." *The Journal of Philosophy* 88 (12) (1991):693–708.

Holton, R. "Deciding to Trust, Coming to Believe." *Australasian Journal of Philosophy* 72 (1) (1994):63–76.

Kelly, Thomas. "The Epistemic Significance of Disagreement." In *Oxford Studies in Epistemology*, edited by Tamar Szabo Gendler and John Hawthorne, 167–96. Oxford: Oxford University Press, 2005.

Kelly, Thomas. "Disagreement, Dogmatism and Belief Polarization." *The Journal of Philosophy* 105 (10) (2008):611–33.

Kelly, Thomas. "Peer Disagreement and Higher-Order Evidence." In *Disagreement*, edited by R. Feldman and Ted Warfield, 111–74. Oxford: Oxford University Press, 2010.

Lackey, Jennifer. "A Justificationists View of Disagreement's Epistemic Significance." In *Social Epistemology*, edited by A. Haddock, A. Millar and D. Pritchard, 298–325. Oxford: Oxford University Press, 2010.

McMyler, Benjamin. *Testimony, Trust and Authority*: Oxford University Press, 2011.

Moran, Richard. "Getting Told and Being Believed." In *The Epistemology of Testimony*, edited by J. Lackey and E. Sosa, 272–306. Oxford: Clarendon Press, 2006.

Shah, Nishi. "A New Argument for Evidentialism." *Philosophical Quarterly* 56 (225) (2006):481–98.

Shapin, S. *A Social History of Truth: Civility and Science in Seventeenth Century England*. Chicago: University of Chicago Press, 1994.

Sosa, Ernest. "The Epistemology of Disagreement." In *Social Epistemology*, edited by A. Haddock, A. Millar, and D. Pritchard, 278–97. Oxford: Oxford University Press, 2010.

Zagzebski, Linda. *Epistemic Authority: A Theory of Trust, Authority, and Autonomy in Belief*. Oxford: Oxford University Press, 2012.

SIX

Disciplines, the Division of Epistemic Labor, and Agency

Fred D'Agostino

According to Goldberg, "the epistemic task has been socially distributed" and, more specifically, there is "a highly differentiated division of intellectual labor."[1]

I think it is important, in developing these ideas, to consider a characteristic situation—that of disciplinary knowledge making—in which "the epistemic task" is increasingly and significantly carried out in societies like ours.[2] While there are certainly distributive and division of labor issues that show themselves in everyday knowledge-making situations (A relies on B's testimony in forming the view that P, where 'P' reports some ordinary state-of-affairs involving medium-sized, readily perceptible objects), "a highly differentiated division of intellectual labor" is perhaps more plainly on display in disciplinary settings.

In this chapter, I want to consider the following matters associated with the social distribution of epistemic agency in disciplinary settings.[3] *First of all,* there is a division of labor in the disciplines; there are many different roles, each with its associated duties and responsibilities that are engaged in discipline-based knowledge making. *Secondly,* I will put under some pressure the standard account of the motivation of epistemic agents working in the disciplines—namely, that they work in order to secure the esteem of their colleagues. I will consider whether, instead, a more satisfactory account is available in terms of the intrinsic rewards associated with the disciplines as practices in Macintyre's sense.[4] *Thirdly,* I will, drawing on a well-known typology of disciplines,[5] consider what kinds of agency it is possible for knowledge-makers to display in the various epistemic situations that distinguish the different disciplines. It will transpire, as I will try to show, that there are limited opportunities for most participants in disciplinary knowledge-making to acquire any significant measure of esteem, on account of the limited forms of agency it is possible for them to exercise. Accordingly, a practiced approach to understanding the motivations of disciplinary agents is to be preferred to an esteem approach.

Before considering these matters, however, it is important to clear the ground when it comes to epistemic agency. In particular, I propose putting to one side the kinds of scruples according to which, as Engel put it, "there can be [no] epistemic agency" "in so far as agency involves acting for a reason."[6] Perhaps this is true when it comes to the matter of A's belief that P; perhaps there is no possibility for agency with respect to belief (because, crudely, believing is not an action). (Cp. Williams 1973.) But, as Engel himself acknowledges:[7]

> That a number of epistemic attitudes involve activity is undeniable: we ... deliberate about whether certain propositions are true, make up our minds about certain topics, take certain things for granted and accept others. We launch inquiries, form hypotheses, reject others and make plans about the theories that we intend to hold.

Indeed we do. Indeed, we do a great many more things besides these, including much more concrete and specific, much more institutionalized activities that are crucially involved in knowledge making in disciplinary settings. It is these activities that I want to describe and assess for their significance. I will do so at right angles, however, without engaging with the metaphysical issues that others might be more interested in and that Engel and Williams are concerned with.

1. ROLES FOR EPISTEMIC AGENTS IN DISCIPLINARY SETTINGS

It will be helpful to say a little about the specifically disciplinary contexts in which knowledge making often now occurs.

An academic discipline has a duplex structure.[8] On the one hand, it is embodied, institutionally, in a geographically dispersed archipelago of university departments at each of which a *collegium* of staff offers instruction in the form of a "major," supervises PhD thesis work, and conducts research. On the other hand, each member of such a department also belongs to a *peer group* of fellow "specialists," again geographically dispersed, who are working on the same topics and deploying the same methods and/or concepts and theories to do so.

Characteristically, undergraduate teaching divides the disciplinary domain more "coarsely" than specialized research activities: while political philosophy might be represented in the undergraduate curriculum as, say, one of a dozen different teachable sub-disciplines, it will show itself, in relation to research activities, in many more specialist fields—more even than the sixty-nine that are represented in the recent *Routledge Companion to Social and Political Philosophy*.[9] Within such a disciplinary setting, specialist knowledge making is itself a highly diverse enterprise, ranging from primary knowledge creation via innovative research activities to, at the other end of a spectrum, knowledge transfer from specialists *qua* teachers to their undergraduate students, as mediated, typically, by textbooks and anthologies which are likely to lag the specialist "research fronts"[10] by anything from a few months (in advanced undergraduate work in the biological sciences) to years or even decades (as in introductory humanities subjects which consider "classic texts" and/or "perennial topics").

Even this brief précis shows several characteristic forms of epistemic agency and implies even more and, indeed, alludes to a division of labor. In particular, we have, already, the roles of researcher, specialist peer or colleague, editor, referee, reader, composer (of textbooks and review articles), teacher, and student.

Even without some further complexities (to be considered shortly), this list is enough to utterly invalidate the notion, admittedly enunciated fifty years ago, that "[i]n the social system of science, . . . we do not find a complex differentiation of roles."[11] There is, instead, and as Goldberg insists, "a highly differentiated division of intellectual labor." Although, for instance, researcher, referee, composer, and student are all part of the same system, the way in which one enacts her role differs from the ways in which the other roles are enacted, and different values and interests are engaged in these different roles. The composer, for instance, is no longer making the judgments about competence and originality that are foremost in the referee's activities, but is focused, instead, on understanding and displaying for others the significance of a variety of original contributions to the "state of play" in a particular specialist area of enquiry. She is synthesizing and contextualizing works whose originality and significance have already been evaluated, e.g. by editors and referees.

Indeed, there is, in this particular value chain, a potentially self-reproducing cycle of activities.

Of course, this representation doesn't begin to capture the complexity of the relationships among the various roles. The cyclic representation may indicate something important about the career of a research article, for instance, passing through the various hands of researcher, referee, and composer before its findings are assimilated to the disciplinary archive and thereafter (for a while anyway) taught to students.[12] And it may also, though less straightforwardly, tell us something about the possible stages in an individual's career. Certainly, the student in some cases becomes the researcher and the researcher in some cases becomes, later in

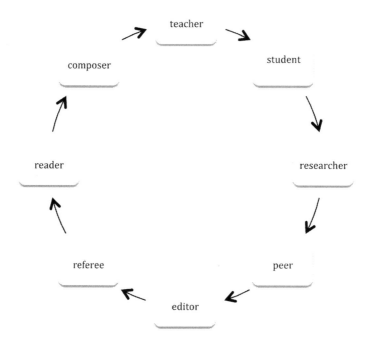

Figure 6.1. The Epistemic Value Chain

her career, the editor or composer. It is misleading, however, in failing to represent the fact that at any given point in time, a particular individual might be both teacher and composer, as well as researcher, peer, editor, referee, and so on. And these facts are not without significance. For what an individual learns as a referee may (indeed should) inform his work as researcher—learning how to assess the work of others, he learns better how to exhibit the virtues and qualities in his own work that he has learned to expect, indeed to demand, when acting as referee. And, similarly, when an individual works, as a composer, to summarize and contextualize recent work in her field, she acquires a perspective, perhaps a broader or deeper perspective, on her own usually more narrowly specialized creative work as a researcher; she better understands, perhaps, the intellectual context of that work.

I said that there were other roles or offices associated with knowledge making in the disciplines that don't, perhaps, fit as neatly into this representation of the value chain. In particular, we have various "coordinating" roles that provide guidance to processes of intellectual development, including *brokers*,[13] or, as Star calls them,[14] "wizards," who are able to direct those researchers working in a particular specialist fields to work in other fields that might be relevant to achieving new insights; *cata-*

lysts,[15] who recognize and confer merit on others, mentoring them to high achievements; and *scholarly leaders and diplomats*,[16] who superintend institutional activities, at various levels and sites including the department, the professional association, the learned academy, and as conference organizers, foundation editors of journals in new fields, and as points of contact with extra-disciplinary stakeholders, ranging from granting bodies and government agencies to the general public.

Again, it is salutary to remember that the humble student might, in the course of her career, come to be a broker or even a diplomat and, as before, that any given individual might at some particular point in his career have many of these "hats" in his closet, shifting between the various roles as needs be and, as already indicated, deriving benefit in discharging one of these roles from his experience discharging others of them.

Indeed, this point, here repeated, is crucial for understanding something about the "highly differentiated division of intellectual labor." In particular, what it shows us is that, potentially, the individual epistemic agent is a microcosm of the disciplinary setting in which she works. Whereas that setting incorporates these various roles (student, diplomat, peer, editor, etc.), any given individual, at least of sufficient seniority, may have played all or many of them during the course of her career and, indeed, might have each of them in her repertoire at some particular time. The division of intellectual labor can be embodied, in other words, both in the community of epistemic agents and, writ small, in some individual members of that community.

2. ESTEEM, RECOGNITION, AND SUSTAINABLE MOTIVATION FOR ROLE PERFORMANCE

One of the most widely accepted hypotheses about the disciplines is that their motive power, if you will, is competition among epistemic agents for esteem and reputations. Whitley's characterization is near-canonical:[17]

> The modern public sciences ... constitute a distinct type of work organization and control in which research is oriented to collective goals and purposes through the pursuit of public scientific reputations among a group of colleague-competitors. In such reputational work organizations, the need to acquire positive reputations from a particular group of practitioners is the main means of controlling what tasks are carried out, how they are carried out, and how performance is evaluated. ... Jobs and resources are allocated largely according to reputations....

On this account,[18] the role of the disciplinary researcher is governed, in effect, by her pursuit of reputational esteem from her specialist peer-group colleagues. That esteem is the coin of the realm when it comes to her career opportunities. The more she acquires, the more likely it is that she will have access to job security, authority within her local departmental collegium and in her dispersed specialist peer group, access to competitive funding, invitations to present her findings at high-profile events or in specially-commissioned collective works, and opportunities to play other roles either as gatekeeper (referee, editor, composer) or as broker or diplomat.

On this account, the prospects for the acquisition of esteem influence all sorts of decisions that the individual member of the disciplinary community has to consider during the course of his career, including, but not exhausted by, these:

- Which research opportunities should I pursue? . . . Those whose payoffs for successful completion will confer the most esteem.[19]
- With whom should I associate within the specialist peer group? . . . Those individuals whose contributions to my own work are most likely to confer esteem, and/or those individuals from whose high esteem I can "borrow" by courtesy.[20]
- How long should I stick with a given research project? . . . So long as the marginal esteem-relevant benefits of the next step in the elaboration of the project exceed the marginal costs of taking that step.[21]

This is an intelligible and internally coherent story about the motive power for epistemic agency in the disciplines. Agents, in whatever role, are motivated by the esteem potential of the various actions and projects that are theoretically open to them.

Of course, as Brennan and Pettit have pointed out,[22] the desire for esteem may most plausibly be construed as a virtual or "standby" motive for the actions of epistemic agents in disciplinary settings and hence it may play no direct role in most of an agent's performances, which may be motivated by other sorts of considerations—perhaps to take on those tasks that are most important for the advancement of knowledge in the specialist field, perhaps simply to exhibit the virtues and conform to the norms which underpin disciplinary or specialist research activities. But the desire for esteem remains in play, on this account, so long as it steers individuals away from activities that, however well motivated on other grounds, do not deliver esteem (in adequate measure).

Certainly, that esteem is (at least sometimes for at least some epistemic agents) a virtual rather than a direct motivation which is more in keeping with the lived experience of epistemic agency in disciplinary settings: many agents experience themselves as problem-solvers, not esteem-pursuers. The Brennan/Pettit proviso therefore represents a valuable corrective to any reductive account of the motivations of epistemic agents. It is nevertheless important to note that it (designedly) does not differ from that reductive account in its implications for courses of behavior by epistemic agents: if an agent is motivated by his commitment to disciplinary norms of responsible epistemic agency to behave in a certain kind of way—to make certain choices of research topics, to stick with or abandon a research project—then, according to Brennan and Pettit, she will do so, subject to various "damping lags" no doubt, only up to the point where those behaviors fail to deliver esteem in adequate measures. Should her norm-compliant behavior cease to deliver esteem, her virtual commitment to the pursuit of esteem will steer her back towards behavior that leads more reliably to the delivery of esteem. Esteem, on this account,[23] "constitutes a standby cause which is ready to kick in if things go amiss and the agent is led away [by conformity to professional norms, for instance] from the behavior" which does reliably deliver esteem. Her behavior, damping lags aside, will be indistinguishable from that of agents who are consciously pursuing esteem.

Despite its coherence and intelligibility, there is, and indeed it has long been known that there is, a significant puzzle associated with this model of disciplinary agency's guiding motives. In particular, that disciplinary agents' decisions are guided by esteem, whether virtually or directly, makes it puzzling that a specialist peer group can continue to recruit and retain as many individuals as those who are indeed recruited and retained. As Gaston already long ago put it,[24] "retention and recruitment would suffer if it appeared that only a few scientists could obtain the recognition that must come after applying one's self to the difficulties and frustrations of research." The key fact, of course, is that the distribution of esteem within most disciplinary communities is heavily skewed;[25] a small number of specialist "stars" receive large measures of esteem, through various media and in various forms, and most participants receive very little, if any, and therefore do not obtain the level of esteem that is supposed, on the canonical story, to motivate their performance as epistemic agents. Why, given this fact, do low-esteem agents continue to participate, or as Fuller puts it,[26] how can "the erroneous agents . . . be allowed to carry on?" Why don't they, as some have wondered,[27] reject as "unfair" their failure to attract esteem in adequate measures? What sustains a skewed (even if not necessarily "unfair") allocation of esteem to agents?

There are several approaches to these questions that do not depend on abandoning the canonical account (of esteem as the primary motivation). These function, if you will, as auxiliary hypotheses that reconcile the undeniable fact of the skewed distribution of esteem with the equally undeniable fact that many unesteemed epistemic agents work very hard in the service of their disciplines.

I will consider two such auxiliary hypotheses, passing over, on this occasion, two others which certainly have merit and will repay further consideration, but which do not fit so well together in reconciling the two considerations which are so starkly in tension with one another ... namely, that (i) disciplinary knowledge-making requires large numbers of agents, whereas (ii) only some of whom will receive esteem for their activities.[28]

Bourdieu suggests a general principle and makes an associated observation that, together, are helpful in understanding how retention and recruitment of disciplinary agents might be sustained even in the presence of a skewed distribution of esteem. The general principle is simple and obvious:[29] "a poor return on scientific investments can lead to accepting or seeking out non-scientific investments of a substitutional or compensatory type." The associated observation trades on the fact that there are numerous non-epistemic roles that are also associated with disciplinary formations. Bourdieu says:[30]

> The plurality of rival principles of hierarchization (which is the basis of the struggles for the imposition of a dominant principle of domination) creates a situation where, as in the field of power overall, each field ... offers innumerable satisfactions which, even when they function as a consolation prize, ... can be experienced as irreplaceable.

Crudely, a discipline can recruit and retain individuals who will not in fact and may indeed predictably not attract the esteem of their colleagues *for their contributions to knowledge making* because there are a variety of other sources of satisfaction available within the disciplinary community: an unproductive (and hence unrewarded) agent can devote himself to academic administration or to committee work.[31] And, indeed, he can serve the specialty or indeed the discipline more generally by his work as editor or even broker, attracting *a different kind of esteem* along the way. If someone finds herself, mid-career, in possession, as it were, of very little of the primary esteem that is associated with the production of disciplinary knowledge, she can still hope to acquire various forms of esteem through other, also worthy activities.

The other, I will call it "ideological," approach to the recruitment and retention difficulties that are posed by skewed distributions of esteem works via "hierarchy-enhancing myths," which, in effect, reconcile the low-esteem agent to his (lowly) position in the disciplinary reputational hierarchy by a congeries of mechanisms, as follows:[32]

> Political sociologists and psychologists have long argued that social ideologies ... often help to ... justify the hierarchical and unequal relationships among groups in society.... Examples of these legitimizing ideologies ... include beliefs in a just world ... in personal causation, ... in personal control, ... in a meritocratic society ... and in the Protestant work ethic.... These beliefs and values legitimize the status quo by holding individuals and groups responsible for their outcomes and by locating the cause of good or poor outcomes within attributes or inputs of that person or group.

On this account, individuals with negligible shares of esteem simply come to accept the legitimacy of their allocations. It is, they think, just that they receive little because, in fact, they have contributed little. They have chosen to participate in a highly competitive domain of activity and, whether through lack of ability or lack of application or just "bad luck," have emerged as losers . . . but, on this account, justly (or at least not unjustly) so.

Obviously, these two mechanisms can complement one another. I may, on ideological grounds, accept my low allocation of reputational esteem and then seek the sorts of substitutes that Bourdieu alludes to. Indeed, this accommodation, through these two mechanisms, is functional for disciplinary activity as a whole. As Crane put it,[33] since most epistemic agents do not attain high esteem (because they have not contributed to knowledge making in ways which earn esteem), "it is necessary to recruit many individuals to an area in order to locate the few who will become highly productive." That they can be recruited and retained, in spite of their poor prospects for attaining high-esteem positions, is therefore reassuring; those that fail will (a) accept that as just, and (b) shift their activities to other, for them more rewarding aspects of disciplinary life.

Seeking to reconcile these undisputed facts—that the distribution of esteem is heavily skewed *and* that large numbers of scholars are recruited and retained in disciplinary settings—we have considered two complementary auxiliary hypotheses and have thereby strengthened the standard account of motivation of epistemic agents: they seek esteem and those who do not obtain it engage either in substitutional activities or in ideological rationalization of their "failure." Notwithstanding the apparent robustness of this by now heavily qualified reputational model of disciplinary agency, its shortcomings on another front, in relation to the lived experience of scholars, suggest it will be worthwhile to consider an alternative.

Writing fifty years ago, long in advance of contemporary interest in epistemic agency in disciplinary contexts, Storer already posed an important (rhetorical) question that points the way forward. He said:[34]

> [I]f the motivational importance of professional recognition lies in its affirmation of successful role-performance, why is it not merited even when an individual has not made an important discovery? Simply obeying the rules of the game, contributing to the general well-being of the community through supporting the norms, would seem to be "successful role-performance."

Indeed. This points us towards another approach to the motivations of epistemic agents working in disciplinary communities, one in which, as Lewis put it (2013, 67), epistemic agents "are not driven purely by external rewards, but also by intrinsic motivation and social preferences . . .

comply[ing] with social and professional norms, and follow[ing] their own curiosity. . . ." On this account, the epistemic agent is a participant, not in a competition for esteem, but, instead, in a practice, in Macintyre's specific sense:[35]

> A practice involves standards of excellence and obedience to rules as well as the achievement of goods. To enter into a practice is to accept the authority of those standards and the inadequacy of my own performance as judged by them. It is to subject my own attitudes, choices, preferences and tastes to standards which currently and partially define the discipline.

As a practitioner, the discipline-based epistemic agent abides by the professional norms and standards current in her disciplinary or specialist community and acknowledges in her work the standards of excellence against which all achievements are to be measured. If her choices are strategically good ones and she is skilled, as well as "obedient," then she may achieve the goods associated with the practice (and accordingly be esteemed by her colleagues). Even if her choices, for any of a variety of reasons, are not strategically sound, she will, however, be recognized as having "played the game," as being a legitimate member of the disciplinary community. And it is, on this account, such *recognition* that is the motive force of the epistemic agent. She counts in her community and is acknowledged by others as counting. While there may be a hierarchy in her community in relation to esteem, each member of that community meets every other member as a fellow participant, as a worthy and respectable member of a community of practice engaged on difficult labors for their own sake. In particular, and taking up a point already made, each participant, including at least some low-esteem participants, will be able to point to her contributions in various secondary roles (editor, broker) that are nevertheless crucial to the sound functioning of the discipline. If she is not esteemed for her contributions to primary modes of knowledge production, she may nevertheless be recognized (and perhaps even, in secondary modes, esteemed) for her performance in other roles in the division of epistemic labor.

In fact, Brennan and Pettit provide a nice account of the distinction between esteem and recognition that I am trading on here and using to interpret Macintyre's account of practices. They say:[36]

> The fact that [esteem] is a comparative attitude means that it is distinct from the equally important attitude that we might describe as giving the person recognition or countenance or standing. . . . Suppose that I give someone esteem or disesteem in a given dimension, I will recognise that person—I will give them respect or countenance—just so far as I treat them as falling within the domain of those who are subject to estimation, positive or negative; I will let the person count. The esteem I give in this sort of case will come in degrees and the degree given will

be sensitive to the comparative performance of relevant others. But the recognition I give will not come in degrees and will not be sensitive in the same way.

It is, on this account, the achievement of recognition (in this specific sense) that the epistemic agent seeks. Sometimes, in seeking recognition, he will also acquire esteem. But, even when he doesn't, he may still be recognizable as a participant, so long as he behaves in accordance with professional norms and standards and orients his work towards the standards of excellence appropriate to his activities.

It is an interesting question whether this model is, as Lewis suggests and many of us would like to believe, more accurate to the lived experience of disciplinary scholars. It does imply, however, a slightly different account of the observed behavior of disciplinary agents: this model, of agents as practitioners, does not imply (as the esteem account does) that, if norm-honoring behavior does not yield (enough) esteem for an agent, then she will behave, instead, in a way that does. In particular, this account is consistent with low-esteem agents continuing to pursue unrewarding research projects in accordance with professional norms and standards (rather than switching to any of the various substitutional activities that Bourdieu enumerates). Notice that this difference shows up even if, as is certainly implied in the standard account, the provision of esteem presupposes norm- and standards-conformity. For even in this case, since conformity is necessary but not sufficient for esteem to be granted, there will, with respect to primary knowledge-making activities, be esteem-losers among the norm-compliant and, on the reputational model, these losers ought to shift their activities to other, potentially more rewarding domains. That they don't always or perhaps even typically switch to secondary activities is therefore prima facie evidence in favor of the recognition model of motivation in disciplinary contexts.

There are, however, other considerations that also seem to put the standard, reputational model under some pressure. To develop these, we will have to consider some of the characteristic differences among the various disciplines. For what the prospects are for the acquisition of esteem vary according to disciplinary situation and some of these situations in fact offer very limited opportunities for esteem to be awarded, and hence put under pressure the idea that esteem is the primary motivator in relation to knowledge making in these situations.

3. SOME CONTEXTS FOR EPISTEMIC AGENCY

On the account developed in the first section of this chapter, there are a variety of different roles that are enacted in the production of knowledge in a disciplinary setting. The researcher, working with her peers, develops new ideas and generates new information. The referee, at the behest of the editor, assesses the cogency, reliability, significance, and originality of the researcher's work.[37] If it passes muster, it will enter into the specialist discursive domain and, in due course, either find an audience (and thereafter be assimilated, perhaps via the work of composers, into the disciplinary archive) or not (in which case the work generates no esteem for the researcher, though she may still get recognition for her efforts).

We have been considering whether the activities of the various role-players are motivated by their desire for esteem or by their fidelity to the norms, standards, and rules of a practice. We have arrived at the conclusion, I think, that the recognition account is more faithful both to the lived experience of scholars and to the fact that scholars seem to persist in primary knowledge-making (rather than substitutional) activities even when they have not succeeded in attracting esteem. There is another reason for suspecting the standard, esteem model, however. In particular, there are quite common disciplinary situations where the problem with pursuing esteem is not that it is an object of competition that is only distributed to the winners, but, rather, that it is, objectively, in short supply, so that its pursuit is not a plausible motive for primary knowledge-making activities in those situations, even for the confidently competitive-minded agent.

By way of concluding, then, I'd like to sketch two different kinds of situations in which knowledge making occurs, and consider both the prospects for esteem and the nature and degree of agency that is available for individuals in them. I will be interested in the degree of autonomy that individuals may be able (and perhaps even required) to exhibit in some, but not all, such disciplinary situations. I will also consider the extent to which esteem (as opposed to recognition) is available for successful performances by agents in these situations.

Whitley developed an influential taxonomy of disciplinary situations, underpinned by some key variables.[38] Whitley considers, for any given discipline or sub-disciplinary research specialty, four key variables, in terms of which he believes the cultural taken-for-granteds of the discipline can be positioned. In particular, we can consider

- how reliant an epistemic agent is on "the specific results, ideas, and procedures of fellow specialists" in presenting her own work as "competent and useful contributions" — *functional dependence*;[39]
- how important it is, in order to achieve recognition or merit the allocation of esteem, "to persuade colleagues of the significance and importance of their problem and approach" — *strategic dependence*;[40]
- how well understood, commonly accepted, and reliable the "work techniques" are — *technical task certainty*;[41] and

- how much consensus there is "about intellectual priorities, the significance of research topics, . . . and the relevance of task outcomes for collective intellectual goals" — *strategic task certainty*.[42]

Variations along these four dimensions define a space of disciplinary types and what is interesting about these types, for our purposes, is that different kinds of agency are afforded in the different disciplinary situations. It is also interesting that, in two common situations, little esteem is reliably on offer, and that its pursuit, even for the self-confident agent, is therefore not plausible as an enterprise.

Where there is high task certainty (both technical and strategic) and high functional dependence, "autonomy is reduced for practitioners."[43] The disciplinary agent working in this situation is enacting a script which already specifies, for him and his colleagues, that they need to attend to each other's work (high functional dependence) and that, in doing so, they need to address the problems or tasks which are already agreed to be important and to use the tools and concepts that are already recognized as reliable and effective. This is, in effect, Kuhn's normal science[44] and it is the site of much routine epistemic activity across a range of different disciplines. This situation is therefore a highly norm-governed one.[45] Whatever the prehistory, which may well have involved the progressive development of a consensus about goals and tools where one did not previously exist, there is now an established agenda for epistemic activity and an agreed set of techniques for prosecuting that agenda. Each individual disciplinary agent simply works to the script that is constituted by the agreed agenda and common toolkit. The form of agency that is available to him is not a very spacious one; in fact, he behaves heteronomously. His behavior is coordinated with that of his fellows in virtue of these commonalities and questions about his competence are reduced to questions about his compliance with a relatively specific set of norms and standards.

One thing that is important about this kind of situation is that while being recognized for competent work is relatively straightforward for most agents, making the kinds of contributions that attract significant degrees of esteem is not. As Whitley put it, "the degree of novelty that can be produced in the modern sciences is restricted by the need to follow collective standards and be relevant to the work of colleagues."[46] And insofar as esteem depends on the production of significant and original work, the degree of esteem that is available for routine performance is also "restricted." This is further acknowledged by Whitley's terminology (1984, 197), in which this type of epistemic situation is characterized as a form of bureaucracy. It is shown, too, in the rich array of facts about citational practices of epistemic agents, in which, in particular, we find that

- most work carried out in this kind of situation is uncited by other practitioners,[47] and so, insofar as citations are a factor in the allocation of esteem, therefore does not confer esteem on them; and
- work that is widely cited in this kind of situation is work that is undertaken during the period when the agenda and the toolkit were being developed, with a sharp fall-off in citations for work later in the development of the normal scientific tradition.[48]

The key point, recalling the Brennan/Pettit discussion, is that esteem is unavoidably comparative and hence requires a scale of significance and/ or originality along which various research performances might be ranked. Where each performance in a particular area is, in effect, a routine enactment of a pre-existing script, there is little to differentiate the various performances and hence a relatively flat "landscape" of possible esteem ratings. However, as Kuhn noted, "scientists [nevertheless] display" "enthusiasm and devotion . . . for the problems of normal research."[49] Since they cannot typically derive (comparatively) high levels of esteem for their efforts in addressing these problems, it is more plausible to suppose that it is recognition (for competent performance) that underpins their activities, and, indeed, their enthusiasm and devotion.

Normal enquiry is one kind of situation in which epistemic agents may be active. On the other hand, where there is low task certainty (both technical and strategic) and low functional dependence, "practitioners . . . develop highly individual research strategies around distinct topics and problems, often with idiosyncratic methods;" "research is rather personal, idiosyncratic, and only weakly coordinated."[50] Here we have an anomic form of disciplinary anarchy with little coordination of research activities among the participants, and, indeed, neither an agenda of topics nor a kitbag of favored tools to serve as a focus of coordination. The individual disciplinary agent is free to pursue her own interests, and in that respect enjoys a high degree of autonomy. But, again, there is potentially not very much esteem, or, indeed even recognition on offer, except insofar as she can persuade her peers that her work is interesting and important, without, of course, being able to rely, in making that case, on already agreed upon (fine-grained) standards and norms.

Here too the recognition model is to be preferred to the esteem model. Of course, some agents working in such anarchic situations will attract a great deal of attention and hence, by whatever measure, a great deal of esteem. But the distribution of esteem will be heavily skewed and the quantities available to most practitioners will be very limited . . . perhaps too limited for the pursuit of esteem to be the most plausible explanation of their "enthusiasm and devotion."

4. CONCLUSION

We have considered a division of epistemic labor that is characteristic of disciplinary knowledge making. We have asked whether esteem or recognition provides a more adequate account of the behavior of individual epistemic agents. Because many agents who are otherwise in good standing disciplinarily do not attain any significant measure of esteem for primary knowledge making activities, it is important to understand what other sources of satisfaction might be available to them. The pursuit of esteem can be complemented by substitutional activities and rationalized, when esteem is not actually obtained, by ideological processes, but the desire for recognition is arguably a more robust motivator. It is also one which survives the observation that, in two characteristic kinds of epistemic endeavor—normal science and customary work in, say, the humanities—there is, anyway, little esteem reliably on offer.

Epistemic agency is exercised in disciplinary circumstances and, accordingly, is subject to disciplinary affordances. Normal science does not afford a great deal of autonomy for individual epistemic agents. Normal humanities research can afford autonomy, but doesn't reliably support the pursuit of esteem.

These are some of the key findings of this enquiry. They show something about epistemic agency in circumstances where there is a division of epistemic labor and, perhaps most importantly, where there are, in fact, a variety of different contexts in which such epistemic agency can be deployed. If agency consists in making choices which are consequential in relation to an individual's values, then there are situations, such as "normal science," where agency is limited, and there are, alternatively, situations in which agency is in fact demanded (because so little is prescribed), but where the values that guide an individual's choice will necessarily be recognition-related, rather than esteem-related.

NOTES

1. Sanford C. Goldberg, chapter 1, section 1.
2. Andrew Abbott, *Chaos of Disciplines* (Chicago: University of Chicago Press, 2001); Tony Becher, *Academic Tribes and Territories: Intellectual Enquiry and the Cultures of Disciplines* (Milton Keynes & Bristol, PA: Society for Research into Higher Education and Open University Press, 1989); Paul Trowler, Murray Saunders, and Veronica Bamber, eds., *Tribes and Territories in the 21st Century: Rethinking the Significance of Disciplines in Higher Education* (London & New York: Routledge, 2012); and Richard Whitley, *The Intellectual and Social Organization of the Sciences* (Oxford: Clarendon Press, 1984).
3. This continues work already reported in Fred D'Agostino, "Disciplinarity and the Growth of Knowledge," *Social Epistemology* 26 (2012): 331–50.
4. Alasdair MacIntyre, *After Virtue: A Study in Moral Theory* (London: Duckworth, 1981).
5. Whitley, *Intellectual and Social Organization*.

6. Pascal Engel, "Is Epistemic Agency Possible?," *Philosophical Issues* 23 (2013): 159.

7. Engel, "Agency," 158.

8. Abbott, *Chaos*, 126; D. W. Hamlyn, *Being a Philosopher: The History of a Practice* (London & New York: Routledge, 1992), 128; Jerry A. Jacobs, *In Defense of Disciplines* (Chicago & London: The University of Chicago Press, 2013), 28, 29; Stephen Toulmin, *Human Understanding* (Princeton: Princeton University Press, 1972), 142, 154; Whitley, *Intellectual and Social Organization*, 57, 81.

9. Gerald Gaus and Fred D'Agostino, eds., *The Routledge Companion to Social and Political Philosophy* (New York & Abingdon: Routledge, 2013).

10. S. Phineas Upham, and Henry Small, "Emerging Research Fronts in Science and Technology: Patterns of New Knowledge Development," *Scientometrics* 83 (2010): 15–38; Keith D. Brouthers, Ram Mudambi, and David M. Reeb, "The Blockbuster Hypothesis: Influencing the Boundaries of Knowledge," *Scientometrics* 90 (2012): 959–82.

11. Norman Storer, *The Social System of Science* (New York: Holt, Rinehart and Winston, 1965), 4.

12. Diana Crane, *Invisible Colleges: Diffusion of Knowledge in Scientific Communities* (Chicago: University of Chicago Press, 1972), 115–16.

13. Ronald S. Burt, "Structural Holes and Good Ideas," *American Journal of Sociology* 110 (2004): 349–99; Etienne Wenger, *Communities of Practice: Learning, Meaning, and Identity* (Cambridge, U.K: Cambridge University Press, 1998).

14. Susan Leigh Star, *Ecologies of Knowledge: Work and Politics in Science and Technology* (Albany: State University of New York Press, 1995), 107.

15. Robert K. Merton, " 'Recognition' and 'Excellence': Instructive Ambiguities," in *The Sociology of Science: Theoretical and Empirical Investigations*, ed. by Norman Storer (Chicago & London: The University of Chicago Press, 1973), 430–31.

16. Warren Hagstrom, *The Scientific Community* (New York & London: Basic Books, Inc., 1965), 44–45.

17. *Intellectual and Social Organization*, 25.

18. Cp. Storer, *Social System*, xxiii: ". . . the institutionally reinforced drive for professional recognition . . . was the energy that would drive the system, . . ."

19. Daryl Chubin and Terence Connolly, "Research Trails and Science Policies," in *Scientific Establishments and Hierarchies*, ed. by Norbert Elias, Herminio Martins and Richard Whitley (Dordrecht & Boston: D. Reidel Pub. Co., 1982), 296.

20. Geoffrey Brennan and Philip Pettit, *The Economy of Esteem: An Essay on Civil and Political Society* (New York: Oxford University Press, 2004), 55ff.

21. Chubin and Connolly, "Research Trails," 296.

22. *Esteem*, 40–48.

23. Brennan and Pettit, *Esteem*, 41.

24. Jerry Gaston, *Originality and Competition in Science: A Study of the British High Energy Physics Community* (Chicago: University of Chicago Press, 1973), 67–68.

25. Keith D. Brouthers, Ram Mudambi, and David M. Reeb, "The Blockbuster Hypothesis: Influencing the Boundaries of Knowledge," *Scientometrics* 90 (2012): 959.

26. Steve Fuller, "A Sense of Epistemic Agency Fit for Social Epistemology," this volume, chapter 2, section 2.

27. E.g. Storer, *Social System*, 150.

28. See Pnina Abir-Am, "Themes, Genres and Orders of Legitimation in the Consolidation of New Scientific Disciplines: Deconstructing the Historiography of Molecular Biology," *History of Science* 23 (1985): 93 for an intriguing third approach, which I call "mythological," and according to which low-esteem participants are to be valued because or insofar as every "glorious discovery [has] depended upon the work of many, more obscure" scholars. Partha Dasgupta and Paul David, "Toward a New Economics of Science," *Research Policy* 23 (1994): esp. 499 give another, economistic account in terms of mixed incentives, with a flat monetary payment associated with teaching, representing, in Fuller's terms (chapter 2), insurance against risk, and a variable component, associated with esteem, earned through risky research.

29. Pierre Bourdieu, *Homo Academicus* (Cambridge: Polity Press, 1988), 99.
30. *Homo Academicus,* 114.
31. See Robert K. Merton, "Priorities in Scientific Discovery," in *The Sociology of Science: Theoretical and Empirical Investigations*, ed. by Norman Storer (Chicago & London: The University of Chicago Press, 1973), 317–18.
32. Brenda Major and Toni Schmader, "Legitimacy and the Construal of Social Disadvantage," in *The Psychology of Legitimacy*, ed. by John Jost and Brenda Major (New York & Cambridge: Cambridge University Press, 2001), 176.
33. *Invisible Colleges,* 35.
34. *Social System,* 26.
35. Alasdair MacIntyre, *After Virtue: A Study in Moral Theory* (London: Duckworth, 1981), 190.
36. *Esteem,* 20.
37. There are, of course, complexities, indeed obscurities, associated with the refereeing process. See, for example, Diana Crane, "The Gatekeepers of Science: Some Factors Affecting the Selection of Articles for Scientific Journals," *The American Sociologist,* 2 (1967): 195–201 and David Shatz, *Peer Review: A critical inquiry.* (Lanham, MD: Rowman & Littlefield, 2004).
38. *Intellectual and Social Organization,* chapter 5.
39. *Intellectual and Social Organization,* 88.
40. *Ibid.*
41. *Op. cit.,* 121.
42. *Op. cit.,* 123.
43. *Op. cit.,* 200.
44. Thomas S. Kuhn, *The Structure of Scientific Revolutions* (Chicago: University of Chicago Press, 1970), chapters III and IV.
45. Whitley, *Intellectual and Social Organization,* 198.
46. *Intellectual and Social Organization,* 12.
47. Brouthers *et al.,* "The Blockbuster Hypothesis," 959.
48. Crane, *Invisible Colleges,* 1–2; Abbott, *Chaos,* 23–24.
49. *Structure,* 36.
50. Whitley, *Intellectual and Social Organization,* 174.

REFERENCES

Abbott, Andrew. *Chaos of Disciplines.* Chicago: University of Chicago Press, 2001.
Abir-Am, Pnina. "Themes, Genres and Orders of Legitimation in the Consolidation of New Scientific Disciplines: Deconstructing the Historiography of Molecular Biology." *History of Science* 23 (1985): 73–117.
Becher, Tony. *Academic Tribes and Territories: Intellectual Enquiry and the Cultures of Disciplines.* Milton Keynes & Bristol, PA: Society for Research into Higher Education and Open University Press, 1989.
Bourdieu, Pierre. *Homo Academicus.* Cambridge: Polity Press, 1988.
Brennan, Geoffrey and Philip Pettit. *The Economy of Esteem: An Essay on Civil and Political Society.* New York: Oxford University Press, 2004.
Brouthers, Keith D., Ram Mudambi, and David M. Reeb. "The Blockbuster Hypothesis: Influencing the Boundaries of Knowledge." *Scientometrics* 90 (2012): 959–82.
Burt, Ronald S. "Structural Holes and Good Ideas." *American Journal of Sociology* 110 (2004): 349–99.
Chubin, Daryl and Terence Connolly. "Research Trails and Science Policies." In *Scientific Establishments and Hierarchies*, edited by Norbert Elias, Herminio Martins, and Richard Whitley. Dordrecht & Boston: D. Reidel Pub. Co., 1982.
Crane, Diana. "The Gatekeepers of Science: Some Factors Affecting the Selection of Articles for Scientific Journals." *The American Sociologist,* 2 (1967): 195–201.

Crane, Diana. *Invisible Colleges: Diffusion of Knowledge in Scientific Communities.* Chicago: University of Chicago Press, 1972.

D'Agostino, Fred. "Disciplinarity and the Growth of Knowledge." *Social Epistemology* 26 (2012): 331–50.

Dasgupta, Partha and Paul David. "Toward a New Economics of Science." *Research Policy* 23 (1994): 487–521.

Engel, Pascal. "Is Epistemic Agency Possible?" *Philosophical Issues* 23 (2013): 158–78.

Fuller, Steve. "A Sense of Epistemic Agency Fit for Social Epistemology." This volume, chapter 2.

Gaston Jerry. *Originality and Competition in Science: A Study of the British High Energy Physics Community.* Chicago: University of Chicago Press, 1973.

Gaus, Gerald and Fred D'Agostino, eds. *The Routledge Companion to Social and Political Philosophy.* New York & Abingdon: Routledge, 2013.

Goldberg, Sanford C. "A Proposed Research Program for Social Epistemology." This volume, chapter 1.

Hagstrom, Warren. *The Scientific Community.* New York & London: Basic Books, Inc., 1965.

Hamlyn, D. W. *Being a Philosopher: The History of a Practice.* London & New York: Routledge, 1992.

Jacobs Jerry A. *In Defense of Disciplines.* Chicago & London: The University of Chicago Press, 2013.

Kuhn, Thomas S. *The Structure of Scientific Revolutions.* Chicago: University of Chicago Press, 1970.

Lewis, J. M. *Academic Governance: Disciplines and Policy.* New York, NY: Routledge, 2013.

MacIntyre, Alasdair. *After Virtue: A Study in Moral Theory.* London: Duckworth, 1981.

Major, Brenda and Toni Schmader. "Legitimacy and the Construal of Social Disadvantage." In *The Psychology of Legitimacy*, edited by John Jost and Brenda Major. New York & Cambridge: Cambridge University Press, 2001.

Merton, Robert K. "Priorities in Scientific Discovery." In *The Sociology of Science: Theoretical and Empirical Investigations*, edited by Norman Storer. Chicago & London: The University of Chicago Press, 1973.

Merton, Robert K. "'Recognition' and 'Excellence': Instructive Ambiguities." In *The Sociology of Science: Theoretical and Empirical Investigations*, edited by Norman

Shatz, David. *Peer Review: A critical inquiry.* Lanham, MD: Rowman & Littlefield, 2004. Storer. Chicago & London: The University of Chicago Press, 1973.

SEVEN

The Distribution of Epistemic Agency

Orestis Palermos and Duncan Pritchard

In this volume, Sanford Goldberg (chapter 1) defines his socio-epistemological research programme by noting that "social epistemology is the systematic study of the epistemic significance of other minds" (section 3).[1] But what can those minds be and how do they differ from the world around us?

Goldberg elaborates by noting that relying on others is not quite the same as relying on the natural world for evidence—as we do, for instance, when we come to know that it's cold outside by seeing someone reaching for their parka or when we discover that we have a mouse problem by finding the droppings under the sink. The difference, explains Goldberg, is that others manifest "*the very results of their own epistemic sensibility*" (chapter 1, section 1).

Goldberg does not employ any of the existing accounts of knowledge and justification in order to offer a detailed definition of what he means by 'epistemic sensibility.' He does venture, however, for a general picture, according to which individuals are members of a 'knowledge environment' that is "structured by various social practices regarding the acquisition, processing, transmission, and assessment of information" (chapter 1, section 1). These social practices, in turn, give rise to certain epistemic expectations between the members of the relevant community, and these expectations can be most profitably thought of as the *norms* that govern the epistemic transactions of that community. Accordingly, it is one's awareness of being answerable to such epistemic norms and expectations that informs one's *'epistemic sensibility'* and promotes one from a mere *epistemic subject* that is capable to possess knowledge to an *epistemic agent* that can participate in the socio-epistemic practices of the relevant community:

> The norms articulate what we properly normatively expect of the relevant individuals as *epistemic agents*: we expect that these individuals have acquired the evidence properly expected of them, and in general that they have behaved with the sort of epistemic responsibility properly expected of them (chapter 1, section 1).

Such norms, Goldberg further notes, may be established explicitly, as a matter of agreement or they may be part of "practices (e.g., information sharing) that emerge over the course of repeated interaction between the parties after the parties mutually (if perhaps only implicitly) acknowledge their mutual reliance on certain aspects of the practice" (chapter 1, section 1).

To illustrate his point, Goldberg gives the following example: Imagine you ask your doctor what's the best treatment for your condition and your doctor replies X. Independently of whether, in order to accept X, you need to merely lack any doubts against the doctor's statement or whether you also need to possess positive reasons in its support, Goldberg notes that part of the story of why you accept X is that, in our society, doctors are "expected to be knowledgeable in certain ways, to be apprised of the best practices, to be responsive to any relevant developments in their specialties, and so forth" (chapter 1, section 1). And it "seems patent that hearers do expect speakers to recognize that when an assertion is made the speaker renders herself answerable to the relevant expectation itself" (chapter 1, section 2).

Accordingly, Goldberg summarizes the hard core of his research programme by noting that "the pursuit of social epistemology is the attempt to come to terms with the epistemic significance of other minds" (chapter 1, section 1). The reason for this is that others "bring their own epistemic sensibility to bear in all sorts of ways as we shape and operate within common epistemic environments" (chapter 1, section 1), and this, Goldberg further explains, is particularly important, because it enables our community's "division of labor to be as far-reaching and systematic as it is" (chapter 1, section 2).

At this juncture, however, Goldberg notes that "it is an open question whether the solitary epistemic subject is the only unit of analysis at which to conduct epistemic assessment" (chapter 1, section 1). If anything, "the development and evaluation of the case for and against [*collective epistemic agents*] ought to be on the agenda of social epistemology" (chapter 1, section 1). The problem for Goldberg, however, is that despite his longstanding interest in social epistemology, he seems to systematically place at the hard core of his programme (or at least in the innermost layers of its protective belt) a strongly individualistic assumption. Specifically, it is highly unlikely that Goldberg would be willing to reject Alvin Goldman's tenet that the "epistemic agents of traditional epistemology are exclusively individuals."[2] Accordingly, Goldberg is unable to address and indeed leaves the question of collective epistemic agency entirely open.

The aim of the present chapter is to extend Goldberg's programme by taking the discussion up from where Goldberg leaves it. To do so, we will elaborate on the idea of *extended knowledge* as we have introduced it in previous work.[3] Extended knowledge is the result of combining virtue reliabilism from mainstream epistemology with active externalism from contemporary philosophy of mind. What is distinctive about this proposal is that it brings into question the assumption that minds in general, and knowledge-conducive abilities in particular, are necessarily restricted within the heads of individuals. Instead, knowledge-conducive cognitive abilities can be occasionally extended to the artifacts we interact with or they may be evenly distributed between several individuals at the same time. If that's correct, then it can have interesting implications with respect to Goldberg's research programme—especially if it can be shown that collective epistemic subjects can qualify as epistemic agents on the basis of being able to *collectively* exhibit an appropriate form of *epistemic sensibility*.

1. EXTENDED KNOWLEDGE

The extended knowledge research programme results from the combination of active externalism from philosophy of mind with virtue reliabilism from contemporary epistemology. Accordingly, before proceeding with our argument for the distribution of epistemic agency, it will be helpful to say a few things about both active externalism and virtue reliabilism, and then explain how the two approaches can be combined. In the process, we will also bring to the fore the way virtue reliabilism understands the notion of 'epistemic sensibility,' which is the cornerstone of Goldberg's framework, and which is going to be central in making the case for epistemic agents that are distributed in nature.

1.1 Active Externalism

As a general approach to the nature of mind, active externalism is standardly contrasted with Hilary Putnam and Tyler Burge's meaning/semantic externalism.[4] Whereas the latter is a passive form of externalism, the former is rather *active*, in that it concentrates on the aspects of the environment that *drive* one's cognitive loops in an ongoing way. Active externalism has appeared in the literature under several labels and formulations—e.g., the extended mind thesis,[5] cognitive integration,[6] environmentalism,[7] location externalism,[8] the hypothesis of extended cognition,[9] the hypothesis of distributed cognition,[10] and so on. For the present purposes, however, we will here concentrate on the latter two formulations.

Focusing on cognitive processing—i.e., any processing that is constitutively involved in the completion of a cognitive task—the hypothesis of extended cognition is the claim that "the actual local operations that realize certain forms of human cognizing include inextricable tangles of feedback, feed=forward, and feed-around loops: loops that promiscuously criss-cross the boundaries of brain, body and world."[11] Similarly, though perhaps more interestingly so, the hypothesis of distributed cognition holds that cognitive processing may not just be extended beyond the agent's head or organism to include artifacts but can be evenly distributed across several individuals along with their epistemic artifacts.[12] Despite its more radical conclusion, however, the hypothesis of distributed cognition differs from the hypothesis of extended cognition only in that cognitive processes and the resultant cognitive systems include not only artifacts but other individuals as well.

With respect to argumentative lines, active externalism—especially in the form of the extended mind thesis—has been traditionally associated with common-sense functionalism.[13] It has been recently argued, however, that contrary to the extended mind thesis, the focus of the extended and distributed cognition hypotheses is not on mental states (such as beliefs and desires, understood in common-sense functionalist terms), but on extended (and distributed) *dynamical* cognitive processes and the overall cognitive *systems* that these processes give rise to.[14] Accordingly, the extended and distributed cognition hypotheses do not need to rely for their support on common-sense functionalism; instead, they can be motivated on the basis of Dynamical Systems Theory (DST), which is perhaps the most powerful, if not the only, mathematical framework for studying the behavior of dynamical systems, in general.[15]

According to this conceptual framework, what is required in order to claim that two (or more) systems give rise to some extended or distributed process and, thereby, to an overall extended or distributed system (either way, to a *coupled* system, in DST terms) is the existence of mutual (i.e., non-linear) relations—that arise out of *feedback loops*—between the contributing parts.[16]

The underlying rationale is that the aforementioned non-linear relations give rise to an overall *integrated* system that consists of all the contributing subcomponents. Typical examples of such systems include two mutually interconnected pendulums, and the watt governor coupled with a rotation engine. In the case of the watt governor and the rotation engine, for instance, their ongoing mutual interaction gives rise to an overall, coupled system that exhibits the distributed property of maintaining a near-constant speed, irrespective of the load or fuel-supply conditions. This is a qualitative property that does not belong to any of the underlying subcomponents but to their ongoing mutual interaction and thus to the overall system as a whole. Accordingly, in order to account for this long-term qualitative behavior, we need to postulate an overall coupled system that consists of both the watt governor and the rotation engine.

In some more detail, in cases like this, there are two main reasons for postulating the overall coupled system: (1) As mentioned just above, the mutual interactions give rise to new systemic properties (like the governor-engine capacity to maintain near-constant speed) that belong only to the overall system and to none of the contributing subsystems alone—therefore, in order to account for such emerging properties, one *has to* postulate the overall extended or distributed system; (2) Said interactions also make it impossible to decompose the two systems' long-term behavior in terms of distinct inputs and outputs from the one subsystem to the other—therefore, in order to make sense of the long-term behavior of the components, one *cannot but* postulate the overall system.[17] For example, in the case of the governor-engine system again, one cannot account for all of the governor's long-term behaviors without simultaneously taking into consideration the behavior of the engine, and *vice versa*. Specifically, it is impossible to decompose all of the governor's behavior in terms of reacting to distinct inputs from, and generating distinct outputs to, the engine, because—when the two of them are mutually interconnected—some of the governor's ongoing behavior both determines and is simultaneously determined by the behavior of the engine (and *vice versa*).

All in all, then, the claim, on the basis of dynamical systems theory, is that in order to have an extended or even distributed cognitive system—as opposed to merely an embedded one[18] —all that is required is that the contributing members (i.e., the relevant cognitive agents and their artifacts) *interact continuously and reciprocally* (on the basis of feedback loops) with each other.[19]

1.2 Virtue Reliabilism

Active externalism holds that cognitive systems are not always bound within the heads of individuals; cognitive systems, instead, may occasionally extend to the artifacts with which individuals interact, or even be distributed amongst several individuals engaging in collaboration. Therefore, in order to introduce active externalism to contemporary epistemology we need an account of knowledge that places at its center the notion of cognitive ability, but in a way that is neutral as to whether cognitive abilities are supposed to be realized within the agent's organismic boundaries or not. And in fact, there is already such an account on offer—*viz.*, virtue reliabilism.[20]

According to virtue reliabilism, knowledge is creditable true belief, where this means that the subject's cognitive success (i.e., getting to the truth of the matter) is attributable to her manifestation of relevant cognitive ability. On this view, cognitive ability is understood as a reliable belief-forming process that has been appropriately integrated into the agent's *cognitive character*, where the agent's cognitive character mainly consists of the agent's cognitive faculties of the brain/central nervous system (CNS), including her natural perceptual faculties, her memory, and the overall doxastic system. In addition, however, it can also consist of "acquired skills of perception and acquired methods of inquiry including those involving highly specialized training or even advanced technology."[21] Here is a relatively weak formulation of virtue reliabilism we can work with, known as 'COGA$_{WEAK}$':[22]

COGA weak
If S knows that p, then S's true belief that p is the product of a reliable belief-forming process, which is appropriately integrated within S's cognitive character such that her cognitive success is to a significant degree creditable to her cognitive agency.[23]

The reason why virtue reliabilists turn to an account of knowledge that stresses the creditable nature of the cognitive success (i.e., believing the truth) as well as its origin in the agent's cognitive ability has to do with the knowledge-undermining epistemic luck involved in Gettier cases. As such cases demonstrate, one's justified belief may turn out to be true without thereby counting as an instance of knowledge. In the typical scenario, one's belief, which is the product of a defective justificatory process, *just happens* to be true for reasons that are extraneous to one's justification. In a lucky turn of events, one's belief, which would otherwise be false (given that it is produced in a defective way), turns out to be true. Contrast this with cases of success through the manifestation of ability. "There is a sense of 'luck' on which lucky success is precisely opposed to success through virtue or ability."[24] When one's true belief is the product of the manifestation of one's ability, then believing the truth cannot have been lucky. Of course, one may still be lucky to believe anything at all (because, say, one could have easily been killed), but believing the *truth* is not lucky. Accordingly, and since credit is normally attributed in cases of success through ability, virtue reliabilists hold that when some agent knows, his belief must be *true because of the manifestation of his cognitive ability*, such that the success be creditable to him.

In other words, virtue reliabilism accentuates the importance of the way one arrives at one's *true* belief—i.e., the *process of getting things right*. It is not enough that one forms one's belief on the basis of virtue (i.e., ability) *and* that one's belief be true: The mere conjunction of these two conditions does not preclude Gettier cases from counting as knowledge. Virtue reliabilists, instead, focus on the *relation* between these two conditions. In order to know, getting to the truth of the matter must be creditable to one and for that to be the case, one's belief must be true because of the manifestation of one's cognitive ability. This is why—and we should mark this to better appreciate the account of collective epistemic agency to follow—virtue reliabilism puts particular weight on the *process* via which one arrives at the *truth* (as opposed to merely believing something that also is, or happens to be, true).[25]

Now, as we mentioned before, according to virtue reliabilism, in order for a belief-forming process to count as a cognitive ability it must be part of the agent's cognitive character. So what could it be required in order for a process to be so integrated? As far as common-sense intuitions are concerned, John Greco has noted that the relevant belief-forming process must be neither strange nor fleeting (i.e., it must be a normal, dispositional cognitive process).[26] Despite such broad intuitions, however, Greco has noted in more recent work that in order for a process to be appropriately integrated into one's cognitive character it must interact cooperatively with it. Specifically he writes: "cognitive integration is a function of cooperation and interaction, or cooperative interaction with other aspects of the cognitive system."[27]

Why does Greco prefer to spell out 'cognitive integration' and 'cognitive character' in this way? The answer has to do with a minimal notion of epistemic responsibility and subjective justification—or, in Goldberg's words, with a weak notion of *'epistemic sensibility.'*[28] Specifically, Greco is after a notion of subjective justification (epistemic responsibility/epistemic sensibility), which is in accordance with epistemic externalism in that it denies that in order to be subjectively justified/responsible/sensible one needs to have access to the reasons for which one's beliefs are reliable.

Unluckily, going into the details of how the integrated nature of one's cognitive character can allow one to be justified in the absence of any positive reasons for one's belief is beyond the scope of the present paper. But the main idea is this: [29] If one's belief-forming process cooperatively interacts with other aspects of one's cognitive system, then it can be continuously monitored in the background such that *if* there is something wrong with it, *then* the agent will be able to notice this and respond appropriately. Otherwise—if the agent has no negative beliefs about his/her belief-forming process—he/she can be subjectively justified in employing the relevant process *by default*, even if he/she has absolutely no positive beliefs as to whether or why it might be reliable. In other words, according to virtue reliabilism, provided that one's belief-forming process is integrated into one's cognitive character such that one would be in a position to be responsive *were there* something wrong (with the process), one can be subjectively justified in holding the resulting beliefs merely by *lacking* any negative reasons against them.[30]

But is this a sense of epistemic sensibility that Goldberg would be happy with? Judging from the following comment, indeed, it seems that he would:

To be sure, we can make efforts to become aware of the various norms that structure our epistemic environment; and we can bring ourselves to reflect self-consciously on the evidence we have for thinking that things in general (or this person or that device in particular) reliably conform(s) to the norms. I surmise that most mature humans do have a good deal of relevant evidence, and that we do on occasion self-consciously reflect in precisely this way. But I submit that we typically do so only when we suspect that the situation doesn't seem right: when the person speaking to us doesn't appear to be fully confident, or is evasive; when one's watch has been making strange sounds recently; when the thermometer reads 20 degrees F, yet we know that it is in the middle of a Chicago summer; and so forth. (Chapter 1, section 1)

1.3 Extended Knowledge

Previously we have argued that reading virtue reliabilism along the lines suggested by the extended cognition hypothesis is not only an available option,[31] but actually necessary for accounting for many instances of knowledge acquired via interacting with epistemic artifacts (e.g., telescopes, microscopes, pen, and paper when trying to solve complex mathematical problems and so on).[32] Moreover, given that the present goal is to demonstrate how the combination of virtue reliabilism and active externalism can allow for epistemically sensible, collective epistemic agents—i.e., groups of individuals that give rise to a collective, knowledge-conducive cognitive ability—we will here limit ourselves to only a few remarks about the strong compatibility between the two views.

To start with, first notice that there is nothing in the formulation of $COGA_{weak}$ or in the concepts involved thereof that restricts knowledge-conducive cognitive abilities to processes within the agent's head. On the contrary, the idea of a cognitive character that may consist of "acquired methods of inquiry including those involving highly specialized training or even advanced technology" seems to be compatible with, or even prefigure, the hypothesis of extended cognition.[33]

This is a good indication for being optimistic about the compatibility of the two views. If we focus on the details of the two theories, however, we can make a much stronger claim. Specifically, both theories put forward the same condition in order for a process to count as part of the agent's cognitive system/character (and, thereby, by the lights of virtue reliabilism, as knowledge-conducive). Just as proponents of extended cognition claim that a cognitive system is integrated when its contributing parts engage in reciprocal interactions (independently of *where* these parts may be located), so Greco claims that cognitive integration of a belief-forming process (be it internal or external) is a matter of cooperative interaction with other parts of the cognitive system.[34]

We see, then, that both in epistemology and philosophy of mind and cognitive science, satisfaction of the same criterion (cooperative interaction with other aspects of the agent's cognitive system) is required for a process to be integrated into an agent's cognitive system and thereby count as knowledge-conducive. It appears therefore that there is no principled theoretical bar disallowing extended belief-forming processes from counting as knowledge-conducive. An agent may extend his cognitive character by incorporating epistemic artifacts to it.

So, for example, in this way we can explain how a subject might come to know the position of a satellite on the basis of a telescope, while holding fast to the idea that knowledge is belief that is true in virtue of cognitive ability. Even though the belief-forming process in virtue of which the subject believes the truth is for the most part external to his organismic cognitive agency, it still counts as one of his cognitive abilities, as it has been appropriately integrated into his cognitive character. Specifically, making observations through a telescope clearly qualifies as a case of cognitive extension as it is a dynamical process that involves ongoing reciprocal interactions between the agent and the artifact. Moving the telescope around, while adjusting the lenses, generates certain effects (e.g., shapes on the lens of the telescope), whose feedback *drives* the ongoing cognitive loops along. Eventually, as the process unfolds, the coupled system of *the agent and his telescope* is able to identify—that is, see—the target satellite. Moreover, the individual subject satisfies $COGA_{weak}$, since his believing the truth is significantly creditable to his cognitive agency (i.e., his organismic cognitive apparatus): It is the subject's organismic cognitive faculties that are first and foremost responsible for the recruitment, sustaining, and monitoring of the extended belief-forming process (i.e., telescopic observation), in virtue of which the truth with respect to the satellite's position is eventually arrived at.

In cases like this, therefore, even though it is the external component that accounts (at least in large part) for the truth-status of the agent's belief, the agent's cognitive agency—i.e., his organismic cognitive faculties—is still significantly creditable for integrating and sustaining the relevant external component into his cognitive system. In other words, in accordance with the demands of $COGA_{weak}$, even though believing the truth is the product of some extended cognitive process, the agent's cognitive success is still significantly creditable to his organismic cognitive faculties. As Andy Clark has expressed the point, "human cognitive processing (sometimes) extends to the environment surrounding the organism. But the organism (and within the organism the brain/CNS) remains the core and currently the most active element. Cognition is organism centered [even] when it is not organism bound."[35]

2. THE DISTRIBUTION OF EPISTEMIC AGENCY

The combination of virtue reliabilism with active externalism, and specifically the hypothesis of extended cognition, can therefore open up the possibility of extended epistemic agents, whose extended cognitive characters can give rise to what we may call 'extended knowledge'—i.e., true beliefs that have been arrived at on the basis of epistemically sensible, extended cognitive abilities. But is it possible to also have 'group knowledge' on the basis of *distributed epistemic agents*? In other words, can collectives exhibit at least the weak form of epistemic sensibility that virtue reliabilism suggests is possible on the basis of the phenomenon of cognitive integration?

To outline our answer, the starting point is to again combine virtue reliabilism with active externalism—though this time in the form of the hypothesis of *distributed cognition*. Doing so can allow for the existence of distributed cognitive characters consisting of a collective cognitive ability that emerges out of the socio-epistemic interactions of the members of a group. Such a collective cognitive ability is going to be irreducible to the sum of the cognitive abilities possessed by the individual members of the group and accounting for it will require the postulation of an overall *epistemic group agent* consisting of all the participating members at the same time.

Of course, before even considering the details of such an answer, one may worry that accounting for collective epistemic agents lacks the requisite motivation—do we really need to postulate such collective epistemic entities? On closer inspection, however, this skeptical attitude is not quite right. To the contrary, postulating epistemic group agents, on the basis of distributed cognition, can in fact allow mainstream epistemology to account for knowledge that is collectively produced and which is thereby distinctively social. Put another way, the idea of epistemic group agents that can act as epistemic subjects in themselves can allow virtue reliabilism to account for the most provocative claim that any social epistemologist could ever make: Namely, that there can be knowledge, which is not possessed by any individual alone but by a group of individuals *as a whole*.

This may indeed sound like a far-fetched possibility but, in fact, there have already been several attempts to introduce this sort of collective knowledge within the literature. Think for example of true beliefs produced by scientific research teams. As several philosophers and ethnographers of science suggest, employing the framework of distributed cognition is the most promising way to analyze such collaboratively produced scientific knowledge.[36]

And this is not the only example. The most thoroughly studied case of group knowledge is the case of what has come to be known within cognitive psychology as *transactive memory systems* (TMSs)[37] —i.e., groups of two or more individuals that collaboratively encode, store, and retrieve information. Consider the following example:

> Suppose we are spending an evening with Rudy and Lulu, a couple married for several years. Lulu is in another room for the moment, and we happen to ask Rudy where they got that wonderful stuffed Canadian goose on the mantle. He says "we were in British Columbia . . .," and then bellows, "Lulu! What was the name of that place where we got the goose?" Lulu returns to the room to say that it was near Kelowna or Penticton—somewhere along lake Okanogan. Rudy says, "Yes, in that area with all the fruit stands." Lulu finally makes the identification: Peachland.[38]

As Wegner et al. explain, during the discussion between Rudy and Lulu the various ideas they exchange lead them through and elicit their individual memories. "In a process of interactive cueing, they move sequentially toward the retrieval of a memory trace, the existence of which is known to both of them. And it is possible that without each other, neither Rudy nor Lulu could have produced the item."[39]

Moreover, and in line with the previous discussion, Barnier et al. suggest that such systems always involve skillful interactive simultaneous coordination between their members.[40] Accordingly, such systems are good candidates for epistemic group agents, because they clearly satisfy the criterion of cognitive integration, as suggested above. As Wegner et al. claim, the members' interaction "gives rise to a knowledge-acquiring, knowledge-holding and knowledge-using system that is greater than the sum of its individual member systems."[41]

So, to give a detailed example of how the suggested framework can account for such group knowledge, consider how $COGA_{weak}$ can account for the way a research team produces knowledge on the basis of an experiment. Even though the knowledge-conducive belief-forming process consists of several experts *and* their experimental devices engaging in reciprocal (socio-epistemic) interactions, the *collective cognitive success* of believing the *truth* of some (scientific) proposition will still be significantly creditable to the group's cognitive agency—i.e., the assembly of the organismic cognitive faculties of its individual members. If anything, it is the assembly of these organismic cognitive faculties that is first and foremost responsible for the emergence and efficient sustaining of the resulting collective's belief-forming process. To paraphrase Clark, *cognition is organism-centered even when it is distributed*. Crucially, however, given that any cognitive success that is collectively produced in this way will only be creditable to the *collection* of the members' cognitive agencies *as a whole* and to none of the individual members alone, it won't be known by any individual alone, but by the group agent as a whole.

Overall, then, by combining an individualistic condition on knowledge, such as $COGA_{weak}$, with the hypothesis of distributed cognition, we can make sense of the claim that p is known by G (the group agent), even though it is not known by any individual alone.

How is this possible and what exactly does it mean? First, we must make clear what it *doesn't* mean. To claim that a proposition p is known by the epistemic group agent as a whole, in the sense presented here, is not to claim that the relevant proposition is collectively known, because it is collectively *believed* (or 'accepted'). This is an alternative approach to collective knowledge that is not necessary to the present approach, and should be clearly distinguished from it.[42] Of course, group knowledge, just as any other type of knowledge, will always involve belief in the proposition known, and so the relevant belief must also, on some appropriate construal, qualify as the belief of the group. But exactly how that idea is to be understood—and there are several live proposals in the literature on this score[43] —is not something that we need to take a stand on here. Indeed, on the present account *any* of these possibilities with respect to group belief may give rise to collective knowledge.

The reason why we do not here need to delve into the details of group belief in order to make the case for collective knowledge has to do with virtue reliabilism's stress on the importance of the *process via which one gets to the truth of the matter*; one's true belief is creditable to one—such that it can thereby constitute knowledge—only if one arrives at the truth in virtue of the belief-forming process (i.e., cognitive ability) one employed to form one's belief. Accordingly, on the basis of virtue reliabilism, which accentuates the importance of the cognitive process via which one arrives at the truth, we can motivate collective knowledge on the basis of cases where arriving at the truth of some matter is the product of a collective belief-forming process. In other words, and to be as clear as possible, on the present *virtue reliabilist* approach to collective knowledge, in claiming that a group can have knowledge that p one is not thereby maintaining that p is collectively and irreducibly believed by that group. Instead, one is rather primarily claiming that the group's getting to the truth of the matter as to whether p could only be *collectively achieved* and is thereby creditable only to the group as a whole.

We therefore see that it is possible to claim that there can be collective knowledge, in the sense that believing the truth can be the product of the cognitive ability of a collective epistemic agent. In order for this claim to also be consistent with Goldberg's research programme (assuming that Goldberg would be willing to allow for epistemic agents that are not strictly individuals), however, we must further clarify the sense in which a collective's cognitive integration allows it to qualify as an *epistemic agent* in its own right. In other words, we need to ask: How can collectives be in a position to exhibit their own epistemic sensibility—at least in the way virtue reliabilism understands the term?

The answer to this final question is that according to virtue reliabilism, epistemic agency is a rather undemanding notion that manifests itself in the initiation, sustenance, and—most importantly—*background* monitoring of the relevant belief-forming process. Specifically, epistemic agency is manifested in the weak and epistemically externalist sense of *epistemic sensibility* that is exhibited by the following conditional: If there is something wrong with the relevant belief-forming process, then the agent will be able to spot this and respond appropriately. Otherwise—if there is nothing wrong—the agent can be *by default* responsible (i.e., subjectively justified) in employing the relevant belief-forming process and its resulting beliefs without even being aware that he does so or that the process is reliable.[44] Accordingly, it is not at all obvious why one should deny *epistemic agency* to a collective cognitive system. After all, it is the assembly of the individual members of the group *as a whole* that initiates and sustains the relevant collective belief-forming process and it is the same assembly operating *as a whole* that is responsible for it. It is the participating members' reciprocal interactions—which bind them together into a unified whole—that allow their cognitive ensemble to effectively be in a position to respond appropriately in cases where there might be something wrong with some part of the overall process.

3. THE DISTRIBUTION OF EPISTEMIC AGENCY AND SOCIAL EPISTEMOLOGY

In conclusion, then, we have seen that according to Goldberg, "social epistemology is the systematic study of the epistemic significance of other minds" (Chapter 1, section 3), where the significance of other minds and the way they differ from the rest of the natural world is that they can exhibit "*the very results of their own epistemic sensibility.*" (3) We have also seen that the combination of virtue reliabilism with the hypothesis of distributed cognition provides a way to expand Goldberg's research programme by incorporating the possibility of collective epistemic agents that can generate group knowledge. Such knowledge is distinctively social because it is not produced by any individual alone but by the members of the relevant group as a whole, and it is only that group as a whole that can count as epistemically sensible with respect to the relevant proposition. Of course, whether this may count as an extension of Goldberg's research programme or as a different research programme altogether is contingent on certain methodological decisions. Specifically, it is an open question whether, alongside his definition of social epistemology, Goldberg would also like to include in the hard core of his research programme the assumptions that (1) epistemic agents do not necessarily have to be individual agents and (2) that virtue reliabilism constitutes at least a necessary condition on knowledge. If (1) and (2) are accepted then the present view can indeed align with Goldberg's research programme in the pursuit of social epistemology. Otherwise, it will constitute a rival to Goldberg's proposal, and the choice between the two (as well as any other existing alternatives) can only be decided on the basis of comparing their future theoretical progress.[45]

NOTES

1. This paper was written as part of the AHRC-funded 'Extended Knowledge' project (AH/J011908/1) which is hosted by the University of Edinburgh's *Eidyn* Philosophical Research Centre, and we are grateful to the AHRC for their support of this research.
2. A. Goldman, "Why Social Epistemology Is Real Epistemology." In Social Epistemology, edited by Haddock, A., Millar, A., and Pritchard, D. (Oxford University Press, 2010), 3.
3. Pritchard, "Cognitive Ability"; Palermos and Pritchard, "Extended Knowledge"; Palermos, "Belief-Forming Processes"; Palermos, "Knowledge and Cognitive Integration."
4. Putnam, "The Meaning"; Burge, "Individualism and psychology."
5. Clark and Chalmers, "The Extended Mind."
6. Menary, *Cognitive Integration*.
7. Rowlands, *The Body in Mind*.
8. Wilson, *Boundaries of the Mind*.
9. Clark and Chalmers, "The Extended Mind."
10. Hutchins, *Cognition in the Wild*; Sutton, "Individual and Collective Memory"; Barnier et al., "The Social Distribution of Agency"; Theiner, Allen and Goldstone, "Recognizing Group Cognition."
11. Clark, "Curing Cognitive Hiccups," section 2.
12. Barnier et al., "The Social Distribution of Agency"; Hutchins, *Cognition in the Wild*; Theiner, Allen and Goldstone, "Recognizing Group Cognition"; Sutton, "Individual and Collective Memory"; Tollefsen and Dale, "Naturalizing Joint Action"; Deborah Tollefsen, "Collective Intentionality."
13. Braddon-Mitchel and Jackson, *Philosophy of Mind and Cognition*. Briefly, common-sense functionalism holds that mental states and processes are just those entities, with just those properties, postulated by our everyday, common-sense, folk psychology.
14. Chemero, *Radical Embodied Cognitive Science*; Palermos, "Loops, Constitution and Cognitive Extension"; Palermos, "Knowledge and Cognitive Integration."
15. For a general introduction to dynamical systems theory see (Abraham, Abraham and Shaw, *A Visual Introduction to Dynamical Systems*). Briefly, *Dynamical Systems Theory* (DST) is the branch of theoretical mathematics that is concerned with the properties of abstract dynamical systems. The general strategy of DST is to conceptualize systems geometrically, in terms of positions, distances, regions, and trajectories within the space of a system's possible states. Overall, it deals with the long-term qualitative behavior of abstract dynamical systems that can represent—and thereby act as models of—concrete dynamical systems such as engines, the atmosphere, our planetary system, our brains, and so on.
16. Barnier et al., "The Social Distribution of Agency"; Tollefsen and Dale, "Naturalizing Joint Action"; Chemero, *Radical Embodied Cognitive Science*; Palermos, "Loops, Constitution and Cognitive Extension"; Theiner, Allen and Goldstone, "Recognizing Group Cognition"; Froese, Gershenson, and Rosenblueth, "The Dynamically Extended Mind"; Wegner, Giuliano, Hertel, "Cognitive interdependence."
17. To preempt a possible worry here, note that the relevant reciprocal interactions need only be continuous during the operation of the relevant coupled cognitive system and the unfolding of any processes related to it. For example, if, as part of her job and during normal working hours, individual S participates in distributed cognitive

system X, S does not need to continuously interact with the other members of X when she is at home. However, whenever X is in operation, S must continuously and reciprocally interact with the rest of the X-members. For a detailed explanation of why the existence of non-linear relations that arise out of reciprocal interactions between agents and their artifacts ensures the existence of extended cognitive systems see Palermos "Loops, Constitution and Cognitive Extension."

18. Adams and Aizawa, *The Bounds of* Cognition; Adams and Aizawa, "The Bounds of Cognition"; Rupert, "Challenges to the Hypothesis of Extended Cognition"; Rupert, *Cognitive Systems and the Extended Mind.*

19. For more details on how dynamical systems theory can help distinguish between the hypothesis of extended cognition and the hypothesis of embedded cognition as well as avoid several other worries with respect to the hypothesis of extended cognition (e.g., the 'cognitive bloat' worry and the 'causal-constitution' fallacy), see Palermos, "Loops. Constitution and Cognitive Extension."

20. See Greco, "Agent Reliabilism"; "Credit for True Belief"; "Ability and the Purpose of Knowledge"; *Achieving Knowledge*; Pritchard, "Cognitive Ability"; Palermos and Pritchard, "Extended Knowledge"; Palermos, "Belief-Forming Processes"; "Knowledge and Cognitive Integration"; "Could Reliability Imply Safety?" and 'Active Externalism, Virtue Reliabilism and Scientific Knowledge." There are several other proponents of virtue reliabilism—most famously Sosa, "Beyond Skepticism"; "Proper Functionalism" and *A Virtue Epistemology*. The reason why only the above references have been included in the main text is to indicate a specific lineage of virtue reliabilism that is particularly apt for our present purposes. In the beginning of the line, however, is Greco, who has, himself, been heavily influenced by Sosa's alternative.

21. Greco, J. "Agent Reliabilism," *Noûs*, 33 (1999): 287.

22. COGA$_{weak}$ stands for Weak COGnitive Agency. This is a weak formulation of virtue reliabilism for two reasons. First, because it is only a necessary condition on knowledge (several epistemologists hold that virtue reliabilism is a necessary component, but to have an adequate theory of knowledge, they argue, it must be further supplemented by either the safety or the sensitivity principle. Second, because, in order to also accommodate testimonial knowledge, it requires that one's cognitive success only be significantly, as opposed to primarily, creditable to one's cognitive agency. Accordingly, 'COGA$_{weak}$' stands for 'weak COGnitive Agency' to indicate that this is an account of knowledge that requires that one's cognitive success be creditable to one's cognitive agency only to a significant (as opposed to primary) degree. For more details on all of these points, see Pritchard, "Cognitive Ability" (cf. Pritchard, "Knowledge and Understanding").

23. D. H. Pritchard, "Cognitive Ability and the Extended Cognition Thesis." *Synthese*, 175 (2010): 136–37.

24. Greco, "Ability and the Purpose of Knowledge," 58.

25. There are problems facing the virtue reliabilist treatment of epistemic luck, but we do not need to get into them here. For further discussion of these issues see Pritchard, "Knowledge and Understanding."

26. Greco, "Agent Reliabilism" and *Achieving Knowledge.*

27. Greco, *Achieving Knowledge,* 152.

28. For an account of the difference between subjective and objective justification, see Carter and Palermos, "The Subjective/Objective Justification Distinction."

29. For a detailed overview of the epistemic internalism/externalism debate and how it maps onto the internalism/externalism debate within philosophy of mind see Carter et al. "Varieties of Externalism." For a detailed analysis of the following minimal yet epistemically adequate notion of subjective justification/epistemic responsibility and its relation to cognitive integration see Palermos, "Knowledge and Cognitive Integration."

30. Palermos, "Knowledge and Cognitive Integration."

31. Pritchard, "Cognitive Ability" and Palermos and Pritchard, "Extended Knowledge."

32. Palermos, "Belief-Forming Processes" and "Knowledge and Cognitive Integration."

33. J. Greco. "Agent Reliabilism," *Noûs* 33 (1999): 287.

34. Elsewhere—see Palermos, "Belief-Forming Processes" and "Knowledge and Cognitive Integration"—it has been argued that both theories put also forward the same broad, common sense functionalist intuitions on what is required from a process to count as a cognitive ability. Briefly, both views state that the process must be (a) normal and reliable, (b) one of the agent's habits/dispositions, and (c) integrated into the rest of the agent's cognitive character/system.

35. Clark, "Curing Cognitive Hiccups," section 9.

36. Giere, "Discussion Note"; Giere, "Scientific Cognition"; Giere, "Agency in Distributed Cognitive Systems"; Giere, "Distributed Cognition"; Knorr-Cetina, *Epistemic Cultures*; Nersessian et al., "A Mixed-Method Approach"; Nersessian et al., "Research Laboratories"; Nersessian, "Interpreting Scientific and Engineering Practices"; Nersessian, "The Cognitive-Cultural Systems"; Thagard, "Societies of Minds"; Thagard, "Mind, Society and the Growth of Knowledge"; Thagard, "Collaborative Knowledge"; Palermos, "Scientific Knowledge."

37. Wegner, Giuliano, and Hertel. "Cognitive interdependence"; Wegner, "Transactive Memory."

38. D. Wegner, T. Giuliano, and P. Hertel. "Cognitive interdependence in close relationships." In Compatible and Incompatible Relationships, edited by Ickes, W. (New York: Springer-Verlag, 1985), 257.

39. Ibid.

40. Barnier et al., "The Social Distribution of Cognition."

41. Wegner, Giuliano and Hertel. "Cognitive interdependence in close relationships," 256.

42. For examples of this alternative approach see Gilbert, "Shared Intention"; "Collective Epistemology"; "Modeling Collective Belief"; "Remarks on Collective Belief"; "Belief and Acceptance" and Tuomela, "Group Knowledge Analyzed."

43. For an overview see Tollefsen, "Collective Intentionality."

44. For more details on this externalist sense of subjective justification see Palermos "Knowledge and Cognitive Integration" and "Could Reliability Imply Safety?"

45. According to Lakatos, "The Methodology of Scientific Research Programmes," 69, the only rational criterion for choosing between two competing research programmes is that one of them can anticipate more theoretically novel facts in its growth.

REFERENCES

Abraham, F. S., Abraham, R. H., and Shaw, C. *A Visual Introduction to Dynamical Systems Theory for Psychology*. Santa Cruz, CA: Aerial Pr, 1990.

Adams, F., & Aizawa, K. *The Bounds of Cognition* (1 edition.). Malden, MA: Wiley-Blackwell, 2010.

Adams, F., & Aizawa, K. "The bounds of cognition." *Philosophical Psychology*, 14 (2001): 43–64.

Barnier, A. J., Sutton, J., Harris, C. B., and Wilson, R. A. "A conceptual and empirical framework for the social distribution of cognition: The case of memory." *Cognitive Systems Research*, 9 (2008): 33–51.

Braddon-Mitchell, D. and Jackson, F. *Philosophy of Mind and Cognition: An Introduction*. Wiley-Blackwell, 2007.

Burge, T. "Individualism and psychology," *Philosophical Review*, 95 (1986): 3–45.

Carter, J. A, Kallestrup, J., Palermos, S. O., and Pritchard, D. "Varieties of Externalism," *Philosophical Issues*, Vol. 24 (2014): 63–109.

Carter, J. A. and Palermos, S. O. "Epistemic Internalism, Content Externalism and the Subjective/Objective Justification Distinction." *American Philosophical Quarterly* (forthcoming).

Chemero, A. *Radical Embodied Cognitive Science.* MIT press, 2009.

Clark, A. and Chalmers, D. "The Extended Mind." *Analysis* 58 (1998): 7–19.

Clark, A. "Curing Cognitive Hiccups: A Defense of the Extended Mind," *The Journal of Philosophy*, 104 (2007): 163–92.

———. *Supersizing the Mind.* Oxford University Press, 2008.

Froese, T., Gershenson, C., and Rosenblueth., D. "The dynamically extended mind." In *Evolutionary Computation (CEC), 2013 IEEE Congress*, pp. 1419–26. IEEE, 2013.

Giere, R. *Explaining Science: A Cognitive Approach.* Chicago: University of Chicago Press, 1988.

———. "Discussion Note: Distributed Cognition in Epistemic Cultures." *Philosophy of Science*, 69, (2002): 637–44.

———. "Scientific Cognition as Distributed Cognition." In *Cognitive Bases of Science*, edited by Carruthers, P., Stitch, S., and Siegal, M. Cambridge: Cambridge University Press, 2002.

———. "The Role of Agency in Distributed Cognitive Systems." *Philosophy of Science*, 73, (2006): 710–19.

———. "Distributed Cognition without Distributed Knowing." *Social Epistemology*, 21, (2007): 313–20.

Gilbert, M. "Shared intention and personal intentions," *Philosophical Studies*, 144 (2009): 167–87.

———. "Collective Epistemology." *Episteme*, 1, (2004): 95–107.

———. "Modeling Collective Belief." *Synthese*, 73, (1987): 185–204.

———. "Remarks on collective belief." *Socializing epistemology: The social dimensions of knowledge* (1994).

———. "Belief and Acceptance as Features of Groups." *Protosociology: An International Journal of Interdisciplinary Research*, 16, (2002): 35–69.

Goldman, A. "Why Social Epistemology Is Real Epistemology." In *Social Epistemology*, edited by Haddock, A., Millar, A., and Pritchard, D., 1–28, Oxford: Oxford University Press, 2010.

Greco, J. "Agent Reliabilism," *Noûs* 33 (1999): 273–96.

———. "Knowledge as Credit for True Belief," in *Intellectual Virtue: Perspectives from Ethics and Epistemology*, edited by DePaul, M. and Zagzebski, L., Oxford: Oxford University Press, 2004.

———. "The Nature of Ability and the Purpose of Knowledge," *Philosophical Issues* 17 (2007): 57–69.

———. *Achieving Knowledge: A Virtue-Theoretic Account of Epistemic Normativity.* Cambridge University Press. 2010.

Hutchins, E. *Cognition in the Wild.* Cambridge: MIT Press, 1995.

Knorr-Cetina, K. *Epistemic Cultures: How the Sciences Make Knowledge.* Harvard University Press, 1999.

Lakatos, I. "Falsification and the Methodology of Scientific Research Programmes." In *Criticism and the Growth of Knowledge*, edited by Lakatos, I. and Musgrave, A. Cambridge University Press, 1970.

Menary, R. *Cognitive Integration: Mind and Cognition Unbound.* Palgrave McMillan, 2007.

Nersessian, N. J., Newstetter, W. C., Kurz-Milcke, E., and Davies, J. "A Mixed-method Approach to Studying Distributed Cognition in Evolving Environments." *Proceeedings of the International Conference on Learning Sciences.* (2003): 307–14.

Nersessian, N. J., Kurz-Milcke, E., Newstetter, W. C., andDavies, J. "Research laboratories as evolving distributed cognitive systems." *Proceedings of The 25th Annual Conference of the Cognitive Science Society.* (2003): 857–62.

Nersessian, N. J. "Interpreting scientific and engineering practices: Integrating the cognitive, social, and cultural dimensions." In *Scientific and Technological Thinking*,

edited by Gorman, M., Tweney, R., Gooding, D., andKincannon, A. 17–56, Erldbaum, 2005.
———. "The Cognitive-Cultural Systems of the Research Laboratory." *Organization Studies*, 27, (2006): 125–45
Palermos, S. O. "Active Externalism, Virtue Reliabilism and Scientific Knowledge." *Synthese*, (forthcoming). DOI: 10.1007/s11229-015-0695-3.
———. "Could Reliability Naturally Imply Safety?" *European Journal of Philosophy*, (forthcoming). DOI: 10.1111/ejop.12046
———. "Loops, Constitution, and Cognitive Extension," *Cognitive Systems Research* 27, (2014): 25–41.
———. "Knowledge and Cognitive integration," *Synthese* 191, (2014): 1931–51.
———. "Belief-Forming Processes, Extended," *Review of Philosophy and Psychology* 2 (2011): 741–65.
Palermos, S. O. and Pritchard, D. H. "Extended Knowledge and Social Epistemology." *Social Epistemology Review and Reply Collective* 2, (2013): 105–20.
Pritchard , D. H. "Knowledge and Understanding," in *The Nature and Value of Knowledge: Three Investigations*, edited by Pritchard, D., Millar, A., and Haddock, A. Oxford: Oxford University Press, 2010.
———. "Cognitive Ability and the Extended Cognition Thesis." *Synthese*, 175 (2010): 133–51.
Putnam, H. "The Meaning of 'Meaning.'" In *Language, Mind and Knowledge*. Edited by Gunderson, K. Minneapolis: University of Minnesota Press, 1975.
Rowlands, M. *The Body in Mind: Understanding Cognitive Processes*. New York: Cambridge University Press, 1999.
Rupert, R. D. 'Challenges to the Hypothesis of Extended Cognition.' *Journal of Philosophy*, 101, (2004): 389–428.
———. *Cognitive Systems and the Extended Mind*. Oxford , New York: OUP USA, 2009.
Sosa, E. "Beyond Skepticism, to the Best of our Knowledge." *Mind*, 97, (1988): 153–88
———. "Proper Functionalism and Virtue Epistemology." *Nous*, 27 (1993): 51–65.
———. *A Virtue Epistemology: Apt Belief and Reflective Knowledge*, Oxford: Clarendon Press, 2007.
Sutton, J. "Between Individual and Collective Memory: Coordination, Interaction, Distribution." *Social Research*, 75 (2008): 23–48.
Thagard, P. Societies of minds: Science as distributed computing. *Studies in History and Philosophy of Science Part A*, 24 (1993): 49–67.
———. Mind, Society, and the Growth of Knowledge. *Philosophy of Science*, 61 (1994): 629–45.
———. Collaborative Knowledge. *Noûs*, 31, (1997), 242–61.
Theiner, G.,Allen, C., andGoldstone, R. "Recognizing Group Cognition." *Cognitive Systems Research*, 11 (2010): 378–95.
Tollefsen, D. andDale, R. "Naturalizing Joint action: A Process-Based Approach," *Philosophical Psychology*, 25 (2011): 385–407.
Tollefsen, D. "Collective Intentionality." *Internet Encyclopedia of Philosophy*, accessed July 19, 2015, http://www.iep.utm.edu/coll-int/.
Tuomela, R. "Group Knowledge Analyzed." *Episteme*, 1 (2004): 109–27.
Wegner, D., Giuliano, T., Hertel, P. "Cognitive interdependence in close relationships." In *Compatible and incompatible relationships*, edited by Ickes, W. 253–76. New York: Springer-Verlag, 1985.
Wegner, D. "Transactive Memory: A Contemporary Analysis of the
Group Mind." In *Theories of Group Behavior*. Edited by Mullen, B. and
Goethals, G. New York: Springer-Verlag, 19856.
Wilson, R. *Boundaries of the Mind: The individual in the Fragile Sciences: Cognition*. New York: Cambridge University Press, 2004.

EIGHT
Toward Fluid Epistemic Agency

Differentiating the Terms Being, Subject, Agent, Person, and Self

Frank Scalambrino

In this chapter I advocate for the importance of the notion of "fluid epistemic agency" for social epistemology. This notion is important because it specifies the difference between two types of epistemic agency. I call the first type "human epistemic agency," and the second type "non-human corporate epistemic agency." Just as rationality is the specific difference between humans and non-human animals, that to which "fluid epistemic agency" refers will hold for humans and not for non-human corporate epistemic agents. Regarding this specific difference it is important for social epistemology that awareness of it brings to light the sense in which the different types of epistemic agency may relate to each other in terms of social classes.

I advance a four part argument in order to advocate for the importance of the notion of "fluid epistemic agency" for social epistemology. The four parts coincide with the following four sections of this chapter. First, because the same terminology may be appropriately used to discuss both types of epistemic agency, I provide terminological clarification in regard to the terms: "being," "subject," "agent," "person," and "self." From out of this clarification I highlight the sense in which both human and non-human corporate epistemic agents may be understood as persons. Second, I discuss the important differences between the two types of epistemic agents in terms of their respective personhood. Third, in light of the difference in their respective senses of personhood, I provide a discussion of how these different types of epistemic agency may relate to each other in terms of social classes. Finally, I conclude the explication of fluid epistemic agency as the specific difference between human and non-human epistemic agency with a discussion of how it is possible for non-human corporate epistemic agents to subject human epistemic agents to a kind of epistemic oppression, i.e. what in the philosophical history of class theory would be called a kind of "hegemonic oppression." This fourth part is particularly important as it shows how fluid epistemic agency applies only to human epistemic agents.

In regard to the important differences between the two types of epistemic agents in terms of their respective personhood, this chapter argues there is an inherent existential dimension at risk for human persons as epistemic agents. However, there is no such risk for non-human corporate persons as epistemic agents. Due to this imbalance of risk the epistemic contributions to a community of knowers from non-human corporate persons may be seen as not only functioning to sustain their presence within a community but also as diminishing the presence and contributions of human epistemic agents within the community, i.e. as perpetrating a kind of hegemonic oppression. It is in this way that human epistemic agents may verge on becoming merely the subjects of non-human corporate epistemic agents.

Therefore, with this chapter I advocate for the importance of the notion of fluid epistemic agency for social epistemology. On the one hand, self-awareness in terms of "fluid epistemic agency" may function as a kind of bulwark against oppression by non-human corporate epistemic agents. On the other hand, this chapter articulates the notion of fluid epistemic agency against the backdrop of the "anchor articles" of this volume. In regard to the question "how does agency need to be understood [for] a conception of normativity to work in the context of social epistemology?" (Fuller, Anchor article section 1), this chapter advocates for speaking of human "beings" as fluid processes conditioning epistemic agency. This may be understood in contradistinction to "other minds" (Goldberg, Anchor article section 1), insofar as what is meant by "mind" in terms of "mind-brain" connection will not hold for some "collective" agents; this is the case even though such collective agents may be non-human epistemic agents contributing significantly to everyday existence in human communities.

Finally, an examination of the relation between the notions of existential risk and epistemic dependence concludes an understanding of fluid epistemic agency as the specific difference between the two types of epistemic agency. Though a complete history of the notion is beyond the scope of this chapter, the background of this examination is informed by the work of Antonio Gramsci and Gilles Deleuze regarding fluid agency. As I will show, because epistemic dependence for humans may function as a condition of becoming subjected to non-human corporate epistemic agents, that to which "existential risk" refers (not personhood or mindedness), differentiates the genus of epistemic agency in such a way that social epistemology may further examine epistemic class theory and the possibility of epistemic oppression by non-human corporate epistemic agents.

1. TOWARD A VOCABULARY OF EPISTEMIC AGENCY: "BEING," "SUBJECT," "AGENT," "PERSON," "SELF"

Understanding the terms "being," "subject," "agent," "person," and "self" as philosophically distinct allows clarification regarding different types of epistemic agency. In regard to social epistemology, then, it is important to clarify these different types of epistemic agency because what *can be* meant in attributing knowledge to an agent is at stake. Moreover, this clarification seems requisite if we are to understand "the appropriate norms for the assessment of acts of reliance (Goldberg, chapter 1, section 1) and "how *agency* need[s] to be understood for a conception of normativity to work in the context of social epistemology (Fuller, chapter 2, section 1).

The term "self" may be used as a pronoun, an adjective, or as a noun. When used as a pronoun, e.g. itself, herself, or himself, it may be thought of as an indexical, i.e. a term that "points" in the general direction of something like the index of a book. When used as an adjective it is understood as indicating identity or sameness, e.g. "belonging to oneself."[1] When used as a noun it may be thought of as referring to a social construct, and interestingly dictionary definitions use language like "person," "character," or "the person someone normally or truly is"[2] to characterize the term.[3] However, it is important to note, the fact that "normal" and "true" may characterize one's self does not determine the nature or extent to which a self is a social construct. In other words, acting more consistently yourself, instead of someone else's self, may be characterized as your "true" self; this is the case even if the self in question was determined primarily by social, cultural, and historical forces, e.g. a true inauthentic self.

The term "person" is used as a noun, and it may refer to "a human being" or a "human individual."[4] Further, as an ethical or moral term it may refer to "a mode of being," and as a legal term it may refer to "one (as a human being, a partnership, or a corporation) that is recognized by law as having rights and duties."[5] One of the ambiguities, then, which needs to be clarified is to what extent, if any, is there a difference between a human person as an epistemic agent and a non-human person as an epistemic agent. The latter, for example, may be understood as a corporation, a collective of humans, or a collective of humans and non-humans (cf. Latour 1999).

The term "agent" is used as a noun, and it may refer to "a person or thing that causes something to happen."[6] Here "thing" is understood as non-person. Hence, "agency" in general may refer to "agent of change," think Aristotle's "efficient cause." Thus "epistemic" agency may be understood as a kind of agency that involves knowledge, i.e. such an agent may be guided by or causing change in regard to knowledge. Therefore, the genus of "epistemic agency" should be further specified into human and non-human corporate epistemic agents. For, in general, non-persons, e.g. non-human forces, may be considered agents of change. However, two primary features of epistemic agency still remain unclear (cf. Scalambrino 2014b). First, insofar as non-human things considered "persons," e.g. corporations and various collectives, may have epistemic status attributed to them, it is not clear how the "personal" agency of such beings differs from the personal agency of human beings. Second, in regard to the "epistemic agency" of such non-human persons the difference must also be examined on the epistemic side. That is to say, the epistemological questions may hinge on an ambiguity regarding different types of epistemic agency.

On the one hand, the ambiguity emerges by leveling the difference in how knowledge may be attributed to human and non-human agents. Though it may be said that there is something of a collective instantiation of knowledge at work in both (e.g. one brain with multiple synapses and one corporation with multiple employees), still clarification is needed in regard to the difference between the types of beings which would then function to sustain the instantiation. On the other hand, agency involves action or causing "something to happen," so in examining the relationship between the agent and the action for which it is supposed to be responsible, the difference between human and non-human epistemic agents should be examined in regard to the "performance" of action.

The term "subject" may be used as a noun, as an adjective, or as a transitive verb. Predictably as a noun it may refer to "the person or thing that is being discussed."[7] As an adjective it suggests being "under the control of a ruler," and as a transitive verb, as in "to subject," it suggests bringing "under control."[8] Regarding subject as a noun, "thing," as non-person, may be understood in a number of ways. Thus, a theme or a topic may be the subject of a discussion; for example, "epistemic agency" as it relates to "social epistemology" may be said to be the subject of this chapter. Subject may also be used ambiguously by referring simultaneously to a subject *as* a theme for discussion. For instance, regarding a discussion titled, "mental agency: a critique of the Cartesian approach to the subject," it is ambiguous whether "subject" refers to the mental, i.e. psychological subject, or mental agency as the subject of the discussion.

As an adjective or transitive verb, "subject" may appropriately refer to the individual elements constituting the instantiation of an agent; however, clarification is needed regarding the type of being subjected to discussion. For example, in a discussion regarding, "unicorns" it may be said that there are various approaches to the subject, because there are different ways the topic may be subjected to exposition. Similarly, the bringing of experience to language may itself subject the process of experience to a linguistic configuration. In this way, from a social perspective, such languaging of experience may be understood as epistemically oppressive. More on this below.

Though the term "being" is the most complicated of the terms discussed thus far, for our purpose here we need not examine all of its complexities. As a noun "being" may refer to "a living thing" or "a state of existing." Further, "being" is understood "primarily [as] the present participial [i.e. participle] form of the verb 'to be,' or simply of the verb 'be.'"[9] In regard to the participle form, for our purpose, it is important to note that when the noun "being" takes on the participle form it represents activity and, thereby, indicates the *agent* of an action, e.g. the biking of a biker. Therefore, when stated with a hyphen, "be-ing" may be understood as referring to a process, e.g. the be-ing of a being or the be-ing of beings.

Such be-ing may be understood as the existential ground for all the other nouns which may appropriately be applied to it. For example the be-ing of a biker is necessary for the biker to *be* biking. In order to emphasize the presence of the existential ground condition of an agent across time, then, the "flowing" of such "across time" is captured by the term "fluid." Hence, before simply understanding a being as determined in terms of the subject of biking, it is possible to examine the process by which the be-ing becomes subjected, i.e. becomes a subject. By thinking in terms of "epistemic fluid agency," then, we need not think of humans solely as subjects. Rather, we may think of humans as dynamic, i.e. fluid, epistemic agents which may be otherwise statically characterized as subjects; in doing so we highlight how such static characterization is always already in terms of some other privileged agent. In this way, the notion of fluid epistemic agency refers to both the radical existential freedom which specifies human epistemic agents as different from non-human epistemic agents and what is at risk in terms of be-ing subjected. This process will be made explicit in the concluding section.

Lastly, we should think of social "subjects," following the work of Gilles Deleuze, along the lines of fluid processes (cf. Deleuze 1992; cf. Deleuze and Guattari 1987; cf. Scalambrino 2015a and 2015d). In this way, as epistemic agents, persons may subject other persons to various transactions or conditions in ways that not only affect persons in regard to their different modes of being but also provide socially relevant self-understandings for the agents (cf. Scalambrino 2014b). That is to say, human be-ing risks subjection in terms of a kind of epistemological oppression. This is important for social epistemology since, on the one hand, in regard to the role of epistemic agents in "the sphere of knowledge production" (Fuller 2000; cf. Fuller, chapter 1,section 1), it involves the fundamental question of social epistemology, i.e. "How should the pursuit of knowledge be organized?" On the other hand, this process of be-ing subjected is also at work in regard to the relational transactions involved in "information-acquisition, -storage, -processing, -assessment, and -transmission" (Goldberg, chapter 1, section 1).

2. FROM DIFFERENT TYPES OF PERSONS TO DIFFERENT TYPES OF EPISTEMIC AGENCY

According to Alvin Goldman, "Knowers are individuals, and knowledge is generated by mental processes and lodged in the mind-brain. Thus . . . it is entirely fitting for epistemology to be concerned with individual knowers and their minds."[10] He then clarifies, "But concentration on the individual to the exclusion of the social is inappropriate."[11] The tendency here may be to understand communities and communities of knowers, i.e. political and civil societies, as including and involving epistemic agents, yet without thinking of the communities as themselves constituting epistemic agents. In contrast, Steve Fuller, in highlighting "the liability model of agency," emphasizes how one's agency may "depend on possibilities that are opened or closed by the presence of other agents" (Fuller, chapter 2, section 2). In other words, both their presence and the interaction of epistemic agents in a community of knowers may open or close possibilities for the community's agents.

Addressing the difference between the types of persons noted above, how the term "person" applies to humans and to corporations, i.e. collective non-human epistemic agents, has significant consequences for epistemic agency. Consider, then, Immanuel Kant's celebrated articulation of human personhood. According to Kant, humans have an original predisposition toward "the good," which may be divided "with respect to function, into three divisions, to be considered as elements in the fixed character and destiny of man."[12] Importantly, Kant's use of the term "predisposition" shows his debt to Aristotle's *Nicomachean Ethics* (cf. 2009, 1105a30–35). For both Aristotle and Kant, in order for a person to be morally excellent the person must be free, i.e. have a free will, have knowledge of the situation in which the action is about to be performed, and have a tendency to choose the type of action which is performed.[13] In other words, a person is not supposed to be able to be morally excellent by accident or in a state of ignorance, and in this way we may say that Kant's "predispositions" resonate with the notion of "habit" in Aristotle.[14]

Kant's three divisions, then, are (1) the predisposition to *animality*, (2) the predisposition to *humanity*, and (3) the predisposition to *personality*, and he characterized their differences in terms of reason and self-love.

Regarding the first of these three divisions, i.e. animality, he suggests "no reason is demanded," and that it may be "brought under the general title of physical and purely *mechanical* [deterministic] self-love."[15] In contrast to the "mechanical" agency of animality, the second and third of the predispositions involve the freedom of human agency. According to Kant, humanity "can be brought under the general title of a self-love which is physical and yet *compares*," and he clarifies "compares" by noting, "we judge ourselves happy or unhappy only by making comparison with others. Out of this self-love springs the inclination *to acquire worth in the opinion of others*."[16] In this way, both humanity and personhood necessarily involve a notion of epistemic agency since the presence of rationality mediates action in terms of knowledge of the situation. However, in contrast to the self-love of animality and humanity, personality involves respect for oneself as a rational agent, which requires an agent's reflective awareness of its rationality beyond humanity's mere application of instrumental reason.

In actualizing the predisposition to personality, through its reflective awareness of its own agency, the agent comes to understand that personal comportment has a rational structure to it. This understanding is twofold. First, the agent realizes the possible actions to be performed may be characterized rationally. Second, the agent realizes its radical existential freedom, since despite the presence of these aspects the agent must still choose, i.e. will, to act. In fact, rationality may become a "sufficient incentive" for the will when the predisposition to personality is actualized. The freedom involved here, for example in regard to rationality as a sufficient incentive, indicates the dignity of the human person and characterizes a kind of morality beyond any ethics based on instrumental rationality. Put another way, respect for oneself as rational and for rationality seems to be a higher development of self-love than that founded on the opinion of others or in comparison with others.

Now, in light of Kant's predispositions, it is important to notice that human epistemic agents can transition across predispositions. Moreover, an agent's tendency toward a predisposition may be understood, following Aristotle, as anchored in its habit structure. As will be discussed below, thinking of Kant's predispositions should help us understand how human epistemic agents can become de-personalized. In other words, if a human epistemic agent's *knowledge* mediating its practical actions can be altered to the point of altering its habit structure, then that epistemic agent may be transitioned "downward" from the personality predisposition, i.e. de-personalized. Further, because it is through knowledge that a human epistemic agent may be transitioned, such a de-personalized agent may be characterized as "epistemically oppressed."

The risk involved here pertains, of course, only to human epistemic agents. Were we to characterize the primary risks for the two different types of epistemic agency under consideration, we might say the primary risk for both types of agent pertains to its survival or its be-ing what it is. In this way, a human epistemic agent's risk is existence-based and the primary risk for a non-human corporate epistemic agent is market-based. In other words, human persons carry a certain existential risk by choosing one understanding of a good life over others with which to regulate their actions. In terms of corporations understood as non-human collective epistemic agents, such agents may be understood as composed of human and non-human agents. Focusing just on the human agents ("employees") which may comprise these corporations, notice how the multiplicity of agents allows for the agency of the corporation to be sustained by radical exchangeability. Hence, whereas their conditions of radical exchangeability and absence of existential risk allow non-human corporate persons to perform actions with machine-like consistency, they also ensure that though such epistemic agents may be thought of as persons, such agents cannot be de-personalized. Despite contributing to a community of knowers as epistemic agents, non-human corporate epistemic agents cannot be de-personalized because they cannot actualize the predisposition to personality.

The imbalance regarding risk may be seen by considering how failure on the part of a non-human corporate person may be remedied with re-branding. What this means is that, even assuming all else is equal, between these different types of persons in terms of their epistemic agency, actions performed in relation to and with consequences for communities of knowers by human persons carry natural moral agency-related consequences which are not at stake for the non-human persons. In other words, even if, following Deleuze, we were to speak of a corporation having a "soul,"[17] corporations do not suffer the same psychic consequences as human epistemic agents, and these psychic consequences are the very indications of moral accountability. Which means further that in the interaction between these two different types of epistemic agents, though a non-human corporate person's actions may influence the predispositions of human persons, i.e. the inclinations of free-will, human persons cannot influence non-human persons in the same way. Hence, human persons could become subjected to a kind of social formation that would serve to sustain the being of the non-human corporate person. In fact, as will be discussed below, to the extent that a non-human corporate person cannot perform epistemic actions without such an eye to self-sustenance, then we may go so far as to say "epistemic oppression" is another name for "marketing" insofar as it relates to corporate sustainability within a community.

3. THE SOCIAL DIMENSION OF EPISTEMIC AGENCY: TOWARD AN EPISTEMIC CLASS THEORY?

The specific difference between the two types of epistemic agency thus far developed may also be articulated in Goldberg's terminology. That is to say, we may understand the imbalance of risk discussed above in terms of "epistemic subjects stand[ing] in various epistemic dependency relations" (Goldberg, Anchor article section 1). Thus two concrete examples may help bring to light the sense in which these different types of epistemic agency may relate to each other in terms of social classes. First, consider how in the process of bringing human experience to language the breadth and depth of experience may become static or solidified around an understanding which originally functioned as "branding" or advertising for a corporation, i.e. "genericized trademarks." Just to name a few: "Zipper," "Zip Code," "Windex," "Xerox," "Coke," "ChapStick," and "Aspirin."

Similarly, consider how Starbucks has changed the way people order coffee. The predisposition to ask for a "Tall," "Grande" or "Venti," instead of small, medium, or large indicates the manner in which a non-human corporate epistemic agent's epistemological "view" of society may influence humans to the point of *habitually* mediating their practical actions. Further, the following, second, concrete example illustrates a different way in which a corporation may participate as an epistemic agent in a community of knowers; recall that former CEO of ENRON, Jeffrey Skilling, was convicted *inter alia* of *conspiracy* and investment *fraud*. The fact that the defrauded funds were not returned evidences the imbalance of risk between the two types of epistemic agency. Just as the scope of the conspiracy and the scale of the fraud were beyond human proportions, Skilling's convictions and punishment, i.e. the punishment of a human for actions conditioned by a non-human corporate person, seemingly could not be commensurate with the kind of just restitution which would have ensued were all parties involved human.

Characterizing the two different types of epistemic agents in terms of class theory may not seem overly controversial when we consider how transhumanists such as Nick Bostrom discuss the difference that superintelligence makes between human and post-human agents (Bostrom 2014). Superintelligence, according to Bostrom, "refers to intellects that greatly outperform the best current human minds across many very general cognitive domains . . . [e.g.] speed superintelligence, collective superintelligence, and quality superintelligence."[18] For, just as, following Latour's discussion noted above, a collective of humans and non-humans (e.g. machines and corporations) may comprise a corporate epistemic agent, according to Bostrom, such collective agents should "have a number of fundamental advantages which will give them overwhelming superiority. [Such that] Biological humans, even if enhanced, will be outclassed."[19] Further, these corporate epistemic agents "may fail to take proper precautions against existential risk."[20] Hence, the transhumanist articulation of different classes of intelligent beings in regard to existential risk is congruent with what I stated above in terms of the imbalance of risk between human and non-human corporate epistemic agents (cf. Scalambrino 2015c).

This speaks directly to the Fuller and Goldberg chapters in regard to risk and trust, respectively (cf. Calabresi and Malamed 1972). On the one hand, in an economic and political field Fuller suggests—along the lines of a "cognitive economics" (cf. Fuller 2015)—"beliefs are seen as akin to investments whose value matures over time" (Fuller Anchor article section 1). On the other hand, Goldberg highlights "the varieties of ways we rely on others" in regard to information exchanged which then informs our epistemic status as epistemic agents (Goldberg Anchor article section 1). Notice, then, conspiring in an economic and political field with epistemic agents "who" are collective non-human persons increases the risk for individual human persons. Hence, from the perspective of a community of knowers in which some of the epistemic agents are non-human corporate persons, epistemic dependence may take on the proportions and force of epistemic oppression, and this may be characterized in terms of an epistemic class theory.

4. TOWARD THE FLUIDITY OF GRAMSCI AND DELEUZE IN THE DIMENSION OF SOCIAL EPISTEMOLOGY

Across the above sections I have argued epistemic dependence may become epistemic oppression for human epistemic agents. In this final section I provide more detail regarding how epistemic dependence may function as a condition by which humans become subjected to non-human corporate epistemic agents. Consider how, as epistemic agents, both human and non-human corporate agents contribute to a community of knowers by performing judgments. When Exxon or Starbucks, for example, change prices or employ "think tanks" to target demographics or when an insurance company employs a celebrity for marketing purposes, these decisions, among others, have efficacious influence in communities of knowers.

This section, then, is important for understanding fluid epistemic agency, not personhood or mindedness, as the specific difference between human and non-human corporate agency. "Personhood" and "mindedness" refer to the Fuller and Goldberg chapters, respectively. That is to say, though the differences between human and non-human corporate epistemic agents may be examined in terms of personhood or mindedness, the specific difference between these two types of epistemic agency, I argue, should be understood in terms of epistemic dependence.

As such, understanding fluid epistemic agency also elucidates the mechanism by which epistemic dependence may become epistemic oppression for human epistemic agents.

Though there are strategies for arguing that personhood and mindedness may refer to both types of epistemic agency, that fluid epistemic agency is the specific difference between these two types of agency means it will apply to human epistemic agents and not non-human corporate epistemic agents. This is precisely the case because there is an imbalance of risk between the two types of agency, and this risk, not personhood or mindedness, places humans at risk for epistemic oppression, i.e. hegemonic oppression in terms of an epistemic class theory. To be sure, there are many ways in which the differences between the two types of epistemic agency may be characterized; however, when we understand human epistemic agency as "fluid epistemic agency," we do so in a way that takes the difference most characteristic of the distinction between the two types, i.e. the specific difference, as a point of departure for thinking through the presence of both types of agency in a community of knowers.

As noted above, "marketing" may be understood as an epistemic contribution to a community of human knowers by non-human corporate epistemic agents, and in light of the imbalance of risk between the two types of epistemic agents discussed in this chapter and continual capitalization (cf. Deleuze 1992; cf. Deleuze and Guattari 1987; cf. Robinson 2005; cf. Scalambrino 1998) and globalization (cf. Jünger 1963) of contemporary cosmopolitan society, the mechanism by which humans may become subjected to non-human corporate epistemic agents is perhaps easier to discern. The following from Gramsci is helpful for understanding how a kind of social formation may take hold of social existence by epistemically influencing how social life is framed by the individual epistemic agents involved.

> A distinction must be made between civil society as understood by Hegel, and as often used in these notes (i.e. in the sense of political and cultural hegemony of a social group over the entire society, as ethical content of the State), and on the other hand civil society in the sense in which it is understood by Catholics, for whom civil society is instead political society of the State, in contrast with the society of family and that of the Church.[21]

Civil society as "political society of the State" suggests for us not only that it is possible but also how to view contemporary society as the constructed product of political forces—as opposed to a collective of the family or Church. Of course, without this distinction to regulate one's meditation, one's situation may be comparable to the young Zen fish who asks of the Master Zen fish, "What is water?" Just as a Church-centric view of society is different from a non-Church-centric view, so too contemporary society may be understood differently by epistemic agents dwelling within "the same" society. In other words, someone with a Church-centric view may still be counted as a member of the "political society of the State."

Further, Gramsci spoke of "the 'spontaneous' consent given by the great masses of the population to the general direction imposed on social life by the dominant fundamental group[s],"[22] and for us the "spontaneity" involved relates directly to existential freedom, and thereby to the imbalance of existential risk between the two types of agency discussed throughout this chapter. First, notice knowledge is involved in the consent of agents which ultimately places them in one determination of political society over another. Second, notice that it is not as if epistemic agents are not "free" within various determinations of political society. Rather, of the extent to which the spontaneity involved eclipses one's ability to act and choose through a different epistemic frame, e.g. a Church-centric versus a non-Church-centric "framing" of society, it is the case that we may say existential freedom is involved (cf. Scalambrino 2015a and 2015c). In other words, when one is carried along by (cf. Heidegger 1962) the direction "imposed on social life," then epistemic dependence becomes epistemic oppression and one's spontaneity is already oriented within a hegemonic frame.

Taking Louis Althusser's famous declaration as a point of departure, i.e. "The ultimate condition of production is therefore the reproduction of the conditions of production"[23] helps contextualize the following from Deleuze.

> Marketing has become the center or the 'soul' of the corporation. We are taught that corporations have a soul, which is the most terrifying news in the world. *The operation of markets is now the instrument of social control* and forms the impudent breed of our masters. Control is short-term and of rapid rates of turnover, but also continuous and without limit.[24]

On the one hand, epistemologically influencing agents may go toward reproducing the conditions of production by way of "repeat customers," e.g. consider the very idea of "brand loyalty." On the other hand, these same epistemological influences are geared toward orienting consumer spontaneity, and, thereby, also function in the service of the "general direction imposed on social life" noted by Gramsci above as an "instrument of social control" noted by Deleuze. Hence, what is at stake may be referred to as "existential freedom," since it neither refers to the freedom within one framing of social life nor another.

How existential freedom is consonant with the notion of fluid epistemic agency may be grasped with greater philosophical acumen by considering how "fluid agency" relates to understanding a "subject" as a "solidified agent," or, put differently, an agent "solidified" by the "subject effect."²⁵ The "subject effect" has a long history in philosophy with its earliest systematization occurring in the work of Aristotle on judgment (cf. Aristotle 2012; cf. Kant 2001, 80). Toward a precise definition of the subject effect, following the language of Husserlian phenomenology, a distinction may be made between "subjectification" and "subjectivation." Whenever a judgment occurs in a situation, the judgment takes the subject-predicate form, e.g. "This is what (ever it is)." Subjectification is indicated by the subject of the statement.

To subject something to judgment is to state its signification in the subject position of a judgment. Such stating is, of course, static, i.e. it treats that which the subjectification signifies as if its identity were apodictic. Notice, we are talking about the identity of the subject, not whether the judgment is apodictic. Once a judgment is performed, it is possible to trace the trajectory back, as it were, in such a way that we may speak of the subject of the enunciation. "Enunciation" is to be understood as another word for "judgment," though judgment seen from this different perspective, i.e. the perspective of performance.²⁶ Insofar as a judgment has been enunciated, then, the "subject effect" refers to the solidification of a "subject of enunciation." That is to say, just like the process of subjectification which occurs by stating some signifier in the subject of a judgment, so too it is as if the stating, i.e. enunciating, the judgment itself silently states an enunciator of the enunciation. Hence, the epistemic assumption or belief in a subject of enunciation resulting from the expression of a judgment is called "the subject effect," i.e. subjectivation. In this way we may say that hegemonic domination does not subject a subject as much as it imposes a market-based horizon, i.e. a "frame," which functions like an enunciation membrane, the subject effect recoil of which constitutes inauthentic political subjects.²⁷

What we need to take away from this brief discussion of "the subject effect" is that just as subjectivation may be understood as conditioning subjectification, the *process* characterized by "existential freedom" conditions both subjectivation and subjectification. That is to say, though the subject of enunciation may be *judged* to be, i.e. understood as, the agent performing the action of judging, the more primordial understanding of agency belongs to the process characterized by "existential freedom." Stated in terms of Kant's predispositions noted above, it is as if when language, instead of moral freedom, determines self-understanding, the tendency of the agent's self-understanding is toward humanity, not the process of personalization. Further, because personhood is sustained epistemically, i.e. through rational mediation of practical action, the process is susceptible to epistemological influence, e.g. the "marketing" process of "suggestion-izing" (cf. Dorpat 1996). It is as if the bringing of experience to language itself subjects the process to a linguistic configuration. Therefore, in contrast to solidified, i.e. subjectified, agency fluid agency refers beyond mere knowledge to a thereby dynamic and free existential dimension; in this way, the specific difference between human and non-human types of epistemic agency may be characterized in terms of "existential risk," i.e. *there is an inherent existential dimension at risk for human epistemic agents which is not at risk for non-human epistemic agents.*

As epistemic agents our actions contribute to a community of knowers *and* constitute that very community. Moreover, because existential risk is an inherent aspect only for human epistemic agents, the very features indicating the imbalance between the two types of agency are integrity and moral accountability. That is to say, it is insufficient to simply acknowledge that community members should be accountable for the constitution of the community and that the concealment of such existential insights may be a consequence of hegemonic domination. Rather, a community of knowers who *know* this may at once also recognize a class struggle between the different types of persons contributing to a community of knowers as epistemic agents. In a Gramsci-inspired turn of phrase, the "collective will" of human persons facing existential risk may be directed toward assembling a (new) "collective intellect-ual," which in terms of social epistemology may be referred to as a collective epistemic agent whose actions, mediated by market-based factors as they may be, restore the value of integrity and the importance of existential accountability into deliberations otherwise confronted by the hegemonic epistemic contributions of non-human corporate persons. Such would be a situation in which integrity and existential accountability permeate the spontaneity of human epistemic agents.

5. CONCLUSION

With this chapter I have advocated for the importance of the notion of "fluid epistemic agency" for social epistemology. I have argued fluid epistemic agency is the specific difference between two types of epistemic agency, i.e. human and non-human corporate.[28] Because the same terminology may be appropriately used to discuss both types of epistemic agency, I devoted one section of this chapter to terminological clarification. That section and its subsequent section showed that neither personhood nor mindedness is to be understood as the specific difference between the two types of epistemic agency. This discussion further brought to light the imbalance regarding existential risk between the two types of epistemic agents. The imbalance was shown to be an inherent aspect of the specific difference between the two types, and this aspect was used to show, from a social perspective, how the different types of epistemic agency may relate to each other in terms of social classes.

Finally, this articulation pertained to the anchor articles especially in regard to Fuller's "liability model of agency" and Goldberg's notion of "epistemic dependence." Hence, this chapter concluded by discussing the mechanism regarding existential risk through which it is possible for non-human corporate epistemic agents to subject human epistemic agents to a kind of epistemic oppression, i.e. a kind of "hegemonic oppression." Whereas human epistemic agents are susceptible to de-personalization through the mechanism characterized as a "subject effect," non-human corporate epistemic agents are not. In this way I have advocated for the importance of the notion of fluid epistemic agency for social epistemology; as a point of departure for thinking in regard to epistemic agency this notion may provide greater philosophical acumen and function as a bulwark against what this chapter has identified as a kind of hegemonic oppression of human persons, by non-human corporate persons, participating in a community of knowers.

NOTES

1. Merriam-Webster.com 2015. http://www.merriam-webster.com. Accessed 06-13-2015.
2. Ibid.
3. I am aware that Carl Jung's specific psychodynamic use of the term "self" has a significant affinity with what I am calling "fluid agency" here; however, it is outside the scope of the current project to develop.
4. Merriam-Webster.com 2015. http://www.merriam-webster.com. Accessed 06-13-2015.
5. Ibid.
6. Ibid.
7. Ibid.
8. Ibid.
9. Oliva Blanchette, *Philosophy of Being* (Washington, DC: The Catholic University of America Press 2003), 13.
10. Alvin Goldman, *Liaisons: Philosophy Meets the Cognitive and Social Sciences* (Cambridge, MA: MIT 1992), 179.
11. Ibid.
12. Kant, Immanuel, *Religion Within the Limits of Reason Alone*, Trans. T. M. Greene and H. H. Hudson. (New York: Harper & Row 1960), 21.
13. Aristotle, 2009 (1110a & 1136b25–27).
14. Frank Scalambrino, "From a phenomenology of the reciprocal nature of habits and values to an understanding of the intersubjective ground of normative social reality," *Phenomenology and Mind*, 6 (2014): 156–67.
15. Ibid, 22–23.
16. Ibid.
17. Gilles Deleuze, "Postscript on the Societies of Control," *October*, 59 (1992), 6.
18. Nick Bostrom, *Superintelligence: Paths, Dangers, Strategies*. (Oxford: Oxford University Press 2014), 52.
19. Ibid.
20. Ibid., 55.
21. Antonio Gramsci, *Quaderni dal carcere, Vol. IV: Passato e presente* (Torino: Giulio Einaudi 2013) 164; quoted in Gramsci 1971, pp. 447–48.
22. Antonio Gramsci, *Selections from the Prison Notebooks*, trans. and ed. Q. Horare and G. N. (Smith London: Lawrence & Wishart 1971), 145.
23. Louis Althusser, *On the Reproduction of Capitalism: Ideology and Ideological State Apparatuses*, Trans. G. M. Goshgarian (London: Verso 2014), 127.
24. Deleuze and Guattari, *A Thousand Plateaus*, 6.

25. This distinction has a deep history in the philosophy of agency (cf. Scalambrino 2013). That is, the distinction here essentially follows the famous difference between "taking hold of one's being in one's own way" (cf. Heidegger 1962) and becoming acclimated to one's cubicle, illustrated in respective vocabularies such that existential therapists speak of "willingness to be a process" and psycho-analysts speak of being better "subjects" (cf. Torrey 1992; cf. Brennan 2003; cf. Sahakian 1975; cf. Scalambrino 2015b). I would like to thank Todd DuBose, Dawn Mitchell, and Vanessa Hicks, as we have had recent conversations regarding existential psychotherapy.

26. Frank Scalambrino, *Non-Being & Memory: A critique of pure difference in Derrida and Deleuze* (Doctoral Dissertation). Retrieved from ProQuest 2011. (UMI: 3466382).

27. Beyond the Heidegger-inspired notion of "inauthenticity," the Deleuze and Guattari-inspired notion of "schizophrenia" may also be operable here.

28. Though a discussion of the extent to which "fluid epistemic agency" may pertain to non-human animals may be interesting, it is beyond the scope of this chapter. Moreover, consider the title of section two of this chapter. Clarification in regard to different types of persons regarding epistemic agency, necessarily excludes animals, as non-persons. Further, though some may say that the relation between non-human corporate epistemic agents and non-human animals may also be characterized in terms of hegemonic oppression, this chapter focuses on humans.

REFERENCES

Althusser, Louis. *On the Reproduction of Capitalism: Ideology and Ideological State Apparatuses*. Translated by G. M. Goshgarian. London: Verso, 2014.

Aristotle. *Nicomachean Ethics*. Translated by Roger Crisp. Oxford: Oxford University Press, 2009.

———. *The Organon: The works of Aristotle on Logic*, edited by E. M. Edghill. Seattle: CreateSpace Independent Publishing Platform, 2012.

Blanchette, Oliva. *Philosophy of Being*. Washington, DC: The Catholic University of America Press, 2003.

Bostrom, Nick. *Superintelligence: Paths, Dangers, Strategies*. Oxford: Oxford University Press. 2014.

Brennan, James F. *History and Systems of Psychology*. Upper Saddle River, New Jersey: Prentice Hall, 2003.

Calabresi, Guido, and Douglas Melamed. "Property Rules, Liability Rules, and Inalienability." *Harvard Law Review*, 85(6) (1972): 1089–28.

Deleuze, Gilles. "Postscript on the Societies of Control." *October*, 59 (1992): 3–7.

Deleuze, Gilles and Felix Guattari. *A Thousand Plateaus (Capitalism & Schizophrenia. Vol. II)*. Translated by B. Massumi. Minneapolis: University of Minnesota Press, 1987.

Dorpat, Theodore L. 1996. *Gaslighting, the Double Whammy, Interrogation, and Other Methods of Covert Control in Psychotherapy and Psychoanalysis*. New York: Jason Aronson.

Fish, Jefferson M. "A University Is Not Walmart." *Psychology Today* (Feb. 03). https://www.psychologytoday.com/blog/looking-in-the-cultural-mirror/201502/university-is-not-walmart. (2015) Accessed 06-24-2015.

Fuller, Steve. *Knowledge Management Foundations*. Woburn MA: Butterworth-Heinemann, 2002.

———. *Knowledge: The Philosophical Quest in History*. London: Routledge, 2015.

Fuller, Steve., and Veronika Lipinska. 2014. *The Proactionary Imperative*. London: Palgrave.

Goldman, Alvin I. *Liaisons: Philosophy Meets the Cognitive and Social Sciences*. Cambridge, MA: MIT, 1992.

Gramsci, Antonio. *Selections from the Prison Notebooks*. Translated and edited by Q. Horare and G. N. Smith London: Lawrence & Wishart, 1971.

———. *Quaderni dal carcere. Vol. IV: Passato e presente*. Torino: Giulio Einaudi, 2013.

Hardt, Michael, and Antonio Negri. *Empire*. Cambridge, MA: Harvard University Press, 2001.

Jünger, Ernst. "Technology as the Mobilization of the World through the *Gestalt* of the Worker." Translated by J. M. Vincent, revised by R. J. Kundell. In *Philosophy and Technology: Readings in the Philosophical Problems of Technology*, edited by C. Mitcham and R. Mackey. 269–89. New York: The Free Press, 1963.

Kant, Immanuel. *Critique of the Power of Judgment*. Translated by Paul Guyer and Eric Matthews. Cambridge: Cambridge University Press, 2001.

———. *Religion Within the Limits of Reason Alone*. Translated by T. M. Greene and H. H. Hudson. New York: Harper & Row, 1960.

Latour, Bruno. "A Collective of Humans and Nonhumans." In *Pandora's Hope: Essays on the Reality of Science Studies*. 198–215. Cambridge, MA: Harvard University Press, 1999.

Merriam-Webster.com. http://www.merriam-webster.com. (2015). Accessed 06-13-2015.

Robinson, Paul. "Pre-emptive War as a Manifestation of Hegemonic Power: Rome, Britain, and the United States." *The Joint Services Conference on Professional Ethics* (JSCOPE). Isme.tamu.edu/JSCOPE05/jscope05.html. (2005). Accessed 06-13-2015.

Sahakian, William S. *History and Systems of Psychology*. New York: John Wiley & Sons, 1975.

Scalambrino, Francesco. *Un uomo sotto la Mole: Biografia di Antonio Gramsci*. Torino: Collana Bancarella, 1998.

Scalambrino, Frank. "A Brief History of the Problem of Agent Causation in the Human and Behavioral Sciences with a Recommendation for Future Research." Paper presented at the 45th Annual Meeting of *Cheiron: The International Society for the History of Behavioral and Social Sciences*, Dallas, Texas, June 21–23, 2013.

———. "From a phenomenology of the reciprocal nature of habits and values to an understanding of the intersubjective ground of normative social reality." *Phenomenology and Mind*, 6 (2014): 156–67.

———. "From a Statement of Its Vision Toward Thinking into the Desire of a Corporate Daimon," *Social Epistemology Review and Reply Collective* 3(10) (2014): 34–39.

———. *Full Throttle Heart: Nietzsche, Beyond Either/Or*. New Philadelphia, OH: The Eleusinian Press, 2015.

———. *Non-Being & Memory: A critique of pure difference in Derrida and Deleuze*. (Doctoral Dissertation). Retrieved from ProQuest. (UMI: 3466382), 2011.

———. Phenomenological Psychology, *Internet Encyclopedia of Philosophy*. http://www.iep.utm.edu/phen-psy/, 2015.

———. "The Vanishing Subject: Becoming Who you Cybernetically Are." In *Social Epistemology and Technology: Toward Increasing Public Self-Awareness Regarding Technological Mediation*, edited by F. Scalambrino. In Press. London: Rowman & Littlefield International, 2015.

———. "What Control? Life at the Limits of Power Expression." In *Social Epistemology and Technology: Toward Increasing Public Self-Awareness Regarding Technological Mediation*, edited by F. Scalambrino. In Press. London: Rowman & Littlefield International, 2015.

Torrey, E. Fuller. *Freudian Fraud: The Malignant Effect of Freud's Theory on American Thought and Culture*. Bethesda, Maryland: Lucas Books, 1992.

NINE

"Epistemic Agency"

A Hegelian Perspective

Angelica Nuzzo

The concept of 'epistemic agency' seems *prima facie* either ambiguous or redundant (or both). And this is the case when the concept is attended to from either side of its components, i.e., alternatively, from the 'epistemic' and from the 'agency' standpoint. On the one hand, it is undeniable that a number of epistemic attitudes are activities or involve some type of activity (such as belief acquisition and revision, the act of making or suspending judgment, proposing guesses, and the like).[1] On the other hand, it is clear that agency always and necessarily presupposes and implies some knowledge on which one acts (such knowledge can be captured by the stronger concept of Aristotelian prudence but can also simply be knowledge of circumstances, projections regarding consequences, calculation of pain/pleasure balances, and the like). Hence the question arises: what is the *specific* issue that the notion of 'epistemic agency' intends to distinctively address beyond these quite obvious ambiguities and redundancies, i.e., beyond the minimal notion of agency implied by some epistemic attitudes, and beyond the minimal notion of knowledge implied by all types of practical activity? Or, to put the point as a challenge to Steve Fuller's general suggestion, is there really a separate "third option" that somehow accounts both for the epistemic character of all forms of agency and for the specific agency that is claimed proper to all epistemic activities?[2] This question, however, should be further qualified as follows: is there a "third option" that is *fundamentally irreducible* to either (traditional) epistemology or moral (or action) theory? Let me sum up at the outset my general position in this regard—the position in support of which I shall begin to argue in the following considerations.

The general sense that I get from the contemporary discussion in which Steve Fuller and Sanford C. Goldberg appear as central protagonists leads me to conclude that the concept of epistemic agency is indeed redundant (and not simply ambiguous). I cannot really see Fuller's alleged "third option" as contributing anything distinctive or irreducible beyond moral theory, at least not in the way Fuller is proposing it here (on his view, the alleged "third option" embraces both sides of the ambiguity or maybe is "even something in between"; it is not, however, something additional or substantially different).[3] And yet, I am willing to explore the possibility that there might be some *philosophically* productive use for the concept of 'epistemic agency.' My suggestion is that if there is, in fact, some philosophically productive use to which the concept can be put, that use must lie in the connection of epistemic agency with the field of social epistemology. In this regard, I think Goldberg does have an important point.[4] The problem then concerns the way in which this connection is achieved. For, just as Fuller's considerations do not seem to leave the realm of moral theory (they seem rather to be set squarely within it including the appeal to champions of moral theory such as Kant and Bentham—although they are additionally sustained by the attempt to conceive the moral-ethical realm in a new, socially expansive way), Goldberg's suggestions remain confined within the epistemological assessment of the relevance of "other minds" branching out to the social sphere from *this* limited position.[5] In contrast to these positions, the claim I want to presently pursue can be summed up in the following way: the concept of epistemic agency may have a philosophical value (unambiguous and non-redundant) only if defined in its basic or primary significance neither in relation to traditional (or analytical) epistemology nor in relation to moral (or action) theory as such but only once it is rooted and re-interpreted within a certain understanding of social epistemology. Importantly, the relation to social epistemology must be primary, and must be instituted at the defining stage since only on this condition may the concept of epistemic agency gain its justification as a possibly un-ambiguous and non-redundant concept. In other words, the concept should not be defined primarily within epistemology or moral theory and *then* re-framed within social epistemology. In order to attain this entry point, however, one needs to mobilize the resources found in some crucial moments of the development of classical German philosophy. This historical detour is necessary if one wants to eliminate the reductive ambiguities that still plague, in my view, the programmatic statements of many social epistemologists including Fuller and Goldberg. In the present considerations I shall limit myself to some passing references to Kant and Fichte and dwell, in particular, on a few crucial aspects of Hegel's philosophy of spirit.[6]

I shall proceed in the following way. First, I bring to the fore and challenge some of the central programmatic points on which Goldberg bases his proposal of an integration of the notion of epistemic agency within social epistemology. While I do generally agree on the relevance of many such points, I believe that they are in need of further elucidation and discussion (much of the ambiguity and redundancy of the concept of epistemic agency hinges on the lack of clarity on these points). Moreover, as they currently stand in Goldberg's proposal these programmatic points do not do enough to integrate socially relevant features in the notion of epistemic agency *as primary and constitutive*. My suggestion is that such primarily social meaning can be gained by bringing into the conversation some of the philosophers of the German classical tradition, Hegel in particular. In a second step of my argument I outline the characters that, in my view, a concept of epistemic agency socially grounded and construed should present. At this stage I address the problem of the link between epistemic agency and social epistemology. And finally, in the last section, I bring in Hegel's theory of spirit as the tool that allows me to flesh out some of the socially relevant characters of the concept of epistemic agency.

1. INDIVIDUALISTIC EPISTEMOLOGY VS. SOCIAL EPISTEMIC ACTIVITY?

On Goldberg's view, one of the central motivations for the turn to 'social epistemology' is the need to move away from the individualism proper to traditional epistemology. My reconstruction of his argument goes, in short, as follows. If, as he claims, the crucial concern of all epistemological theories, i.e., of any theory of knowledge, is the concern with the "sources" of knowledge, the fact that there are "distinctly social sources of knowledge" requires the development of an epistemological theory that is in the position to appreciate the "distinctly" social practices of knowledge acquisition. Now, since traditional epistemology is on Goldberg's view constitutively "individualistic in its orientation,"[7] i.e., is centered on the individual subject (which must be taken to mean: is centered on the individual as the "subject" of knowledge and as the "source" of knowledge), this discipline is constitutively unable of appreciating the "distinctly" social character of the epistemic activity insofar as such character escapes the activity of epistemic individual subjects and is irreducible to individual sources of knowledge (this, I take it, must be the meaning of that repeated "distinctly"). Let me pause here and underline a few problematic points in this argument.

First of all, what shall we understand by "sources" of knowledge? Is it simply meant, *objectively*, the passive, material repository of information available as an empirical given in the world (the natural, human and non-human, social, historical world) from which knowledge (subjectively) obtains (in form, among others, of inference)? Or is it meant something closer to what a transcendental and constructivist perspective such as Kant's would mean by it,[8] in which case the sources of knowledge are, *subjectively* (in the sense of belonging properly, i.e., subjectively, to reason), the *active* cognitive powers that *actively* determine the way in which the material for knowledge is taken up and processed precisely into knowledge? This clarification— roughly, objective vs. subjective meaning of the "sources" of knowledge—is crucial if we are to successfully move (i) from the issue of the individual sources of knowledge to the issue of the allegedly "distinctly social" sources of knowledge; but also if we are to successfully move (ii) from the issue of the *subject* of knowledge to the type of *agency* that underlies epistemological processes—a move, this latter, that despite being so crucial remains unclear and difficult to follow in Goldberg's proposal (and just as murky in Fuller's). In other words, what is it that makes a source of knowledge alternatively individual (even individualistic) or social; what is it that makes it "distinctly" so; and what is the relation between the *source* of knowledge and the epistemic *activity* (and not just the subject) at stake in knowledge?

In a transcendental perspective such as the Kantian it is clear that the source of knowledge is that from which knowledge actively issues, hence is an a priori activity (a *Handlung* of the mind in the most proper sense). From this position the transition to the concept of epistemic agency seems to follow without too much difficulty (and yet, even in this case, the distinction between epistemic subject and epistemic agent constitutes an important problem in its own right). Fichte has radicalized this Kantian notion of agency, which remains, I submit, an important notion of epistemic agency transcendentally and not socially grounded. In fact, on Kant's premises, there is nothing intrinsically social to reason's cognitive activity. If one were to argue for a constitutively social form of epistemic agency on this transcendental basis, then one would have to prove that there are irreducibly social forms of activity (a priori sources in the subjective sense) from which a distinct type of knowledge issues. Fichte may indeed be seen as taking this route with his notion of intersubjectivity; and there is a sense in which Fichte does open the way for Hegel's dialectical conception of spirit. But this is certainly not Goldberg's line of argument, as he does not endorse the transcendental perspective in the first place.

On the other hand, if the "sources" of knowledge are understood objectively, we should distinguish two different cases. At this juncture, it appears, we may place the contemporary emergence of the problem of intersubjectivity or interpersonality raised by Fichte—and now, perhaps, proposed by Goldberg. The problem regards the conditions under which a philosophical theory (be it an epistemological or a moral theory) can make room for a distinct concept of intersubjectivity as irreducible to the atomic individuality under which the (thinking and acting) subject is conceived. There is certainly the reductive possibility entailed in Goldberg's observation that in the "individualistic framework" of traditional epistemology "others" (or "other people" in their "antics and appearances") are indeed present but only with the status of "*evidence* from which one can come to know things through inference."[9] On this view, other people have exactly the same epistemic status (or are empirical epistemic sources in the same way) as all other things in the world. Indeed, if the problem of intersubjectivity is the problem of the theoretical justification of other human beings, the issue does not properly subsist within an empiricist perspective (or, with Fichte, in the perspective of the "egoist" and, with Goldberg, of individualistic epistemology). Herein intersubjectivity hardly needs justification. Take Locke's case, for example. On his view, we recognize another human being by simply discerning an analogy with ourselves—an analogy that plays itself out first and foremost at the physical level. "I think I may be confident, that whoever should see a creature of his own shape or make, though it had no more reason all its life than a cat or a parrot, would call him still a man; or whoever should hear a cat or a parrot discourse, reason, and philosophize, would call or think it nothing but a cat or a parrot; and say, the one was a dull irrational man, and the other a very intelligent rational parrot."[10] On a merely empirical plane, in the lecture on the *Scholar's Vocation* (1794), Fichte ascertains that experience teaches us "that the representation of rational beings outside of ourselves is contained in our empirical consciousness." In fact, the theoretical justification of the assumption of other rational human beings outside of ourselves is not an empirical problem.

At the empirical level individualism—and even the solipsism of the "egoist"—seems to be the only possibility. The issue that divides the transcendental philosopher and the theoretical egoist (or, the individualist epistemologist opposed by Goldberg) is precisely "whether this representation [of another rational being] corresponds to something outside of it."[11] Ultimately, Fichte argues, the problem of other rational being is a specifically *transcendental* problem. And it is, more precisely, a *practical* problem—a moral and a juridical problem that rests on a transcendental epistemological basis. Fichte's novel discussion of this issue—a first attempt, I submit, at a social epistemology—takes place in the *Foundations of Natural Right* (1794–1795). It is important to stress, however, that if there is in Fichte a question regarding other rational being as a specifically intersubjective *source* of knowledge he thinks that this question rests on the solution of the preliminary problem concerning the justification of the existence of other beings as ourselves.[12]

But there is also another possibility of reading the objectivity of the sources of knowledge, this time in a non-reductive sense. We can admit the existence of *distinctively social*—and more generally of intersubjective—sources of knowledge according to which other people are not just things that function as sources of evidence for inferences, and nonetheless we can still hold on to an individualistic view of the subject of knowledge. In other words, it is still possible to think that knowledge acquisition and practices are "often and perhaps even typically"[13] social activities and yet to posit that the agent of such activity remains an atomic individual. As I will claim in the last section, this is the case of the individual actor of Hegel's "civil society." This case, however, despite its individualism remains an important moment of a fundamentally *social* epistemology such as the one proposed by Hegel at the level of his theory of ethical life (*Sittlichkeit*). This indicates that individualistic and social orientations are not mutually exclusive epistemological perspectives, as Goldberg seems to maintain. It follows that the charge of individualism—or the notion that an individualistically oriented epistemology cannot account for the social dimension of knowledge acquisition—is not enough to warrant the move to a specifically social epistemology within which the notion of epistemic agency ought to be developed. To put this point differently, the opposition between individualistic and social orientation is not very helpful when at stake is the problem of defining the concept of epistemic agency. For, such opposition is not the true point of contention when at stake is the evaluation of the *sources* of knowledge; and is not the true point of contention when at stake is the evaluation of the *practices* and *processes* of knowledge acquisition. The consequence is that if the concept of epistemic agency is to be linked to the field of social epistemology so as to be constitutive of the social character of the agency involved in epistemic processes, it has to be linked in a way that recognizes that individual and social subjects, agents, and processes are not mutually exclusive or antinomic but are instead mutually mediated or dialectically and dynamically interconnected.

Goldberg sums up the task of social epistemology by defining it as "the attempt to come to terms with the epistemic significance of other minds." By which he means precisely that other people are not just "sources" of knowledge (as I take it, in the objective, passive sense) like all other things and events in the world but are themselves "epistemic subjects in their own right" who are, in addition, *loci* of epistemic agency of their own. Thereby Goldberg anchors the concept of epistemic agency in the field of social epistemology. There are two broad and interconnected issues that should be raised with regard to this programmatic claim—issues the clarification of which is crucial if one is to decide whether the gesture of grounding epistemic agency within social epistemology in a non-ambiguous and non-redundant way is successful or not. On the one hand, at stake is the claim regarding the "epistemic significance of other minds."[14] This seems to me an insufficient characterization of the distinctive topic of a social epistemology within which room is to be made for a non-ambiguous and non-redundant concept of epistemic agency. And the problem is compounded by the fact that the preliminary question—metaphysical (and not only epistemological)—of how I am to distinguish between things and minds in the first place (minds to which agency is to be additionally ascribed) looms large in the background of Goldberg's argument without ever being explicitly raised. And yet, the epistemic significance of other mind may be connected to or may even be dependent on the specific way in which I first come to identify—metaphysically, epistemically, practically, morally—other minds *as minds* and not as things. This is the preliminary question raised transcendentally by Fichte in the discussion mentioned above. Now it seems to me that Goldberg's central issue concerning the epistemic significance of other minds can hardly be addressed without raising this question. But the problem regards, secondly, the very relevance or significance of those other minds: can their relevance be only epistemic when those other minds are seen as endowed of an agency of their own—i.e., when others are not just subjects of cognitive practices but specifically agents? On what basis do I get to ascribe agency to other minds? Herein we touch on the second dichotomy that, in addition to the one between individualism and social orientation, if not mediated seems to condemn the concept of epistemic agency to an inevitable redundancy. This is the opposition between theoretical or epistemological and practical or moral concerns. What we have, yet again, is the ambiguity mentioned at the outset to which Fuller explicitly refers. How do we get from the recognition of the epistemic significance of other minds to the ascription of agency to them—of an agency that is not just practical (or moral-practical) but, in turn, epistemic? But at stake is also the connected central task of distinguishing between epistemic "subjects" and epistemic "agents." What does the latter notion add to the former? In this regard I do not find any clarity in the discussion—neither in Goldberg nor in Fuller.[15]

2. EPISTEMIC AGENCY: EPISTEMIC PROCESSES AND EPISTEMIC AGENTS

I shall move on now to a brief outline of the notion of epistemic agency which in the next section I begin to develop in the framework of Hegel's theory of spirit. As I wish to conceive it, the concept of epistemic agency drives (traditional) epistemology away from the issue of qualifying and differentiating the *sources* of knowledge, and brings the epistemological concern to bear instead on the very *activity* of knowledge production and practices. It is on this basis that epistemology then gains its fundamental extension to the sphere of the social. I consider this sense of epistemic agency as following from Kant's (and Fichte's) transcendental attention to the subjective (and, with Fichte, properly intersubjective) sources of cognition once the apriorism of the transcendental framework is abandoned in favor of a developmental, historically determined account of cognitive processes such as Hegel's. While on Kant's theory the epistemic subject in its cognitive powers is engaged in the *Handlung* of knowing (the "I think" of the original unity of apperception is the transcendental *locus* of subjectivity or is subject that acts but is not properly agent as this, for Kant, is only the moral-practical agent), for Hegel the epistemic subject becomes directly agent. Significantly, this transformation on the epistemological side is accompanied by a transformation in the meaning of the practical whereby the strictly "moral" sense that the practical has in Kant's theory (as formally connected to the law of freedom) yields to the broader social (and political juridical historical) meaning that it displays in Hegel's philosophy of "objective spirit."

What needs to be understood now, in the first place, is this shift from the transcendental epistemic subject—Kant's individual and Fichte's intersubjective subject—to the epistemic agent that is "spirit" in its dialectical constitution.[16] This shift is made possible by a radical gesture, namely, by the gesture that turns away from subjectivity *tout court* in order to concentrate on the immanent unfolding of the process or the action itself. At its foundational or purely logical stage (which is developed by Hegel throughout the *Science of Logic* as the first discipline of his philosophical system) this action takes place, as it were, without a subject (or an agent) doing the thinking and the knowing at stake in the logical development.[17] What is at stake in Hegel's logic is, paradoxically or rather dialectically, a process without a presupposed subject undergoing it—a process that unfolds without (or independently of) a metaphysical subject or *res cogitans*, without (or independently of) a transcendental subject such as Kant's (or Fichte's) "I think," without (or independently of) a phenomenological consciousness-subject such as the one Hegel himself had mobilized in the early *Phenomenology of Spirit* (1807). Hegel's logical account of the immanent development of thinking and knowing does not depend on—and is not shaped by—a presupposed subject doing the thinking, i.e., does not depend on the subject's metaphysical, epistemological, psychological, historical position or constitution. It is instead such as to generate as its last result the very structures of "subjectivity" that form the basis of the reality of "spirit" (*Geist*). Spirit is subjectivity that displays a dynamic, developmental reality: spirit is nothing besides or above its actual manifestations in the productions of knowledge and in different forms of practical activity. Spirit is Hegel's replacement of the metaphysical substance of early modern metaphysics but also the successor of Kant's transcendental "I think." It is precisely on this logical basis that spirit successively develops—or properly 'realizes'—its manifold forms of reality. Spirit manifests itself as individual and as collective, social spirit—as "subjective" and "objective" but also as "absolute" spirit. While these forms as systematically successive (in the progression of the *Encyclopedia*, for example), they are dialectically co-present and not antinomically opposed within spirit's actuality. The latter displays the dynamical intertwinement of cognitive and practical activities, of individual and social manifestations. In sum, I propose viewing Hegel's logic as a 'dialectic epistemology without subject' which constitutes the systematic basis for a 'social epistemology of spirit.' The concept of epistemic agency is placed at the point of intersection—or as the bridge—mediating between these two.

The idea of *agency* stresses the fact that at stake are epistemic *processes*. Truth is not a final static achievement that can be considered—and displays its value—independently of the process that leads to it and constitutes it. On Hegel's view truth is famously "the whole," and the whole is the dynamic movement or the process of its immanent genesis and development.[18] Accordingly, Hegel's dialectic epistemology is interested in the *dynamic* constitution of truth as a complete and systematic *process*; it is interested in the epistemic process itself, not in isolated epistemic claims or beliefs.[19] As I propose it here, the idea of epistemic agency captures this fundamental concern of Hegel's dialectic. Now the process that truth itself is unfolds according to its own immanent laws and norms, independently of any presupposed subject in the sense that it does not owe its structure and dynamic character to the nature of the subject that carries it out (to its being human reason, for example, or to its being sensibly affected or socially and historically determined in a certain way). For, it is rather the very process of truth that eventually, in its conclusion, constitutes the subject as the agent that in the end can ascribe knowledge and beliefs to itself as its own. Hegel's logic (and his epistemology of spirit) presents the nature of different types of action (processes of knowledge acquisition and transmission as well as forms of practical activity) not on the basis of the nature of the subject that performs such action but by immanently considering the dynamic of action itself—i.e., on the basis of internal features proper to action such as, for example, the different ways in which contradiction is dealt with and eventually overcome, the way in which difference and otherness are either disruptive for or integrated in the results to be attained, and the like. It follows that it is the epistemic subject—or properly agent—that depends on the epistemic process—or on the agency displayed in action—and not the other way around. Along with truth as a concluded process, the agent is co-constituted as the result of the epistemic activity, and is positioned within it accordingly. On Hegel's premises, then, the epistemic agent is the subject to which, retrospectively, a certain epistemic activity (or set of epistemic activities) can be ascribed as her own (or who ascribes them to herself as her own). The agent is constituted by the epistemic activity as the determined agent that she is, and not vice versa. It is not the case that the epistemic process is that which a certain epistemic subject/agent, in a certain epistemic position—individual or social, in possession of certain epistemic virtues as opposed to others—is able to perform. The opposite is rather the case: it is the fundamentally and constitutively 'individual' or, alternatively, 'social' dynamic of the epistemic process that makes the epistemic agent respectively into an 'individual' or into a 'collective' social agent. To put this point differently, the agent is not a reality presupposed once and for all in its constitution (in its capacities, virtues, and values) and is not therefore taken for granted. Rather, the agent—its capacities, virtues, values—is that which it accomplishes in the process or the activity that manifests that agent in the world. Accordingly, practices are not 'social' because enacted by an agent that is declared 'social'—which only shifts the problem of determining what counts as a 'social' agent. Practices are 'social' on the basis of the very structure (and logic) of the action involved so that social practices (in this sense) constitute social agents as the only agents to which those practices can be ascribed.

It is clear then that within this Hegelian framework a concept of epistemic agency can be proposed that stands as the alternative to the notion of epistemic agency of virtue epistemology advocated, for example, by Catherine Elgin (but following, more broadly, from the Aristotelian model). While in Elgin's case the epistemic status of the agent's beliefs "derives from the excellence of her epistemic character," on Hegel's model "the agent's epistemic excellence" derives from the process in which "epistemically estimable beliefs" are attained.[20] It is also clear that with this formulation we have reached one of the central demonstrative tasks of a Hegelian 'social epistemology of spirit' that employs the concept of epistemic agency as one of its main operative concepts. In connection with our previous discussion of Goldberg's proposal, we can conclude that one of the foremost tasks of such a Hegelian social epistemology consists in indicating what it is, in the very structure of epistemic processes, that qualifies them as either individual or social or both, i.e., as social but individually centered.

3. EPISTEMIC AGENCY AND HEGELIAN GEIST

In this last step of my argument I shall turn directly to Hegel's philosophy of spirit in order to offer a general illustration and some examples of the two crucial points I made in the previous sections. Let me sum them up. (i) First, in discussing Goldberg's proposal of a notion of epistemic agency linked to the program of social epistemology, I brought to the forefront the necessity of overcoming the redundancy of the concept of epistemic agency by overcoming (in the strict sense of Hegel's *Aufhebung*) the antinomic opposition or alternatively the ambiguous merging into each other of the epistemic and moral-practical, the individual and social dimensions of subjectivity and agency. (ii) Second, I have proposed a notion of epistemic agency that takes the dynamic structures of action (or agency) as primary and foundational, and defines or first institutes the determinate character of the epistemic agent on its basis. This, I have suggested, is the epistemic agency proper to Hegelian spirit in contrast to the Aristotelian view of action or to the epistemic agency of virtue epistemology. Now I want to flesh out these two points with the support of Hegel's texts.

(i) To the first issue first. In its logical foundation, which systematically accounts for its agency, Hegelian spirit is a dynamic reality dialectically placed beyond the antinomic opposition of theoretical and practical, individual and collective forms of action. Hegel's philosophy of spirit in the *Encyclopedia* (1830) is famously divided into the three systematically successive spheres of "subjective," "objective," and "absolute" spirit.[21] It is important to stress, however, that the systematic succession implies the relation of dialectical *Aufhebung*. The stage that is achieved last in the developmental movement of spirit is the broadest reality that contains in itself the forms from which it has arisen and that it has overcome; but it is also, in point of fact, the more developed reality that encompassing the previous stages makes them possible for what they are. For spirit, the act of leaving contradictory forms behind is the very act of integrating them within one's own reality as constitutive. Now this general dialectical relation should be brought to bear on the connection between the epistemic and moral validity, the individual and the social character of spiritual processes. The result is a dialectical configuration of the relation between individualist and social epistemology that is quite different than the one advanced by Goldberg, and offers, at the same time, a dialectical configuration of the relation between epistemic and practical validity that can dispel the ambiguity and redundancy displayed by the concept of epistemic agency in both Goldberg's and Fuller's sense.

In the psychology of "subjective spirit" Hegel explores the fundamental structures and activities that first constitute *Geist* in its subjectivity and individuality—an individuality that is both self-centered and related only to itself, atomistic and isolationist, as it were. What we have at this level is the exploration of basic as well as more advanced cognitive and practical processes of the mind, i.e., the fundamental ingredients of what Goldberg identifies as traditional individualistic epistemology. "Theoretical spirit" is spirit involved in the processes of cognition (in the processes of intuition, representation, abstraction, generalization, linguistic expression and symbolization, etc.), which eventually lead to its liberation from the giveness of the world and to the transition to "practical spirit." To put this point in terms of our present discussion: the general epistemic activity that constitutes spirit to "free intelligence" eventually discloses the practical validity of that very epistemic agency. The epistemic agency proper to "theoretical spirit" dialectically transforms it into "practical spirit."[22] At the level of practical spirit, however, the world is not simply a world of given things and external objects (with Goldberg, to be used as evidence for inferences) but is a world in which the will relates to other free subjective wills—to other spiritual individuals acting according to freedom (and willing the realization of freedom) and recognizing them as such. And yet, at the end of subjective spirit, the will is still individualistic and atomistic. There is no collective action that is willed or performed, there is no collective reality to which the individual belongs in order to perform its action. To put this point with Hegel's famous claim in the *Phenomenology*, the will is an "I" not a "We."[23]

The epistemic activity of theoretical spirit is the progressive appropriation of external data that seem to be simply and immediately "found"[24] in the world. This is the activity whereby spirit makes what it finds into its own (into its own "being" and "property") by appropriating it cognitively (in intuition and representation and ultimately in conceptual form).[25] This general epistemic activity is successively articulated into different functions, i.e., develops through the epistemic processes that to a different degree succeed in appropriating the givenness of raw data by conferring to it cognitive validity, i.e., certainty and truth value. The epistemic agency of "theoretical spirit" is accordingly developed into an analysis of intuition, perception, representation (to which belong the functions of memory, imagination, language, and sign production in general), and conceptual thought. Hegel's general point is that the epistemic act of transforming pre-existing data and raw material (*Stoff*)[26] into meaningful cognitive contents is the practical act of liberation whereby spirit is constituted, for the first time, into a *free* individual. In other words, by appropriating a pre-existing independent content cognitively, knowledge renders the individual free from the external dependency on the world as that which was initially only given (external data and raw material) becomes something produced and owned by the subject (a cognitive content). It is clear then that theoretical intelligence and practical will are two distinct functions of subjective spirit, which are instituted as different stages in the dialectical development of the same 'epistemic agency' that is spirit's self-liberation to individual subject. In other words, and back to our discussion, what we have here is the articulation of the 'epistemic agency' that first institutes spirit in its psychological subjectivity and individuality. There is no redundancy or ambiguity between the epistemic and the practical moment here but a dialectical relation within a dynamic process.

Moreover, if we look at the systematic placement of this result, whereby spirit's epistemic agency constitutes its reality as properly "practical spirit,"[27] we can draw a second important conclusion. In instituting theoretical intelligence into "practical spirit," subjective spirit's epistemic agency attains the first station in the "path in which the will makes itself into *objective* spirit."[28] This means that spirit's individual (and individualistic) agency is the first step toward its fundamental 'socialization.' Again, there is no irreconcilable alternative between spirit's individualism and its social reality. Rather, Hegel's point is that it is only having used up all the resources of individualistic epistemology that spirit is brought to recognize that there are subjects in the world that are spiritual agents (and not simply objects or things to be used only as mere sources of evidence, as Goldberg puts it). On the basis of the dialectical relation of *Aufhebung*, the individualistic epistemology of subjective spirit remains at the foundation of the process of spirit's socialization but is also, at the same time, overcome and expanded within a radically different sphere of spirit's reality, namely, at the level of "objective spirit." For, it is only here, and more precisely at the level of "ethical life" (*Sittlichkeit*), that "true freedom" finds its ground. In other words, social epistemology is grounded on an individualistic epistemology; it is not its radical alternative. In addition, on the basis of the reversal proper to dialectical *Aufhebung*, we must also recognize that individualistic epistemology is justified only within the broader framework of a social epistemology of spirit.

Now freedom, for Hegel, is a form of spiritual activity that on the ground of the movement of subjective spirit just sketched out has both an epistemic and a practical dimension. Hegel's claim is that truly free agency necessarily exceeds the individual self-interested and self-centered subjectivity explored in the sphere of subjective spirit, and produces a different 'agent' (or rather, the different plural agents) that is the social, collective, and institutional reality emerging in the realm of "ethical life." In other words, Hegel's suggestion is that the individual free spirit that is the result of the movement of subjective spirit is not the agent that both epistemically and practically can carry out truly free agency. The latter can only be the social agent that arises in the sphere of "objective spirit." In this sphere, "true freedom" is social, i.e., on Hegel's view, implies that "the will has as its ends not a subjective, self-interested content but a universal content."[29] This is accordingly the first condition of all epistemic agency that displays a social character.

Within the sphere of ethical life, at the level of "civil society," which encompasses the world of economic relations and transactions among atomic self-interested individuals, we find an important example of the intersection and coordination of individual and social epistemic agency, but also of the dialectical reversal whereby an individualistic epistemology is justified and fulfilled on the basis of a social epistemology. The agent of civil society is a particular, "concrete person" characterized by a totality of needs, natural feelings, and arbitrary volitions. However, it is only through her relation to other individuals that she is able to fulfill her volitions and satisfy her needs. This interaction is the basis of the "universality" that characterizes this utterly individualistic sphere.[30] Although individual ends are "selfish," based on merely personal interests and motivations, they are also social and intersubjectively mediated. We have here a case of epistemic agency that is both radically individualistic and socially mediated and achieved. First, individual ends are conditioned by the universal context of reciprocal interaction because this context alone allows for those ends to be realized. Subsistence, welfare, and rights of the individual are interwoven with and dependent on the subsistence, welfare, and rights of all.[31] This is the argument that makes political economy an accurate description of individual self-interested action within civil society. But Hegel offers a second reason for the "universality" of individual action in this sphere, which displays the social character of individual epistemic agency. Within civil society selfish motivations and practical feelings are acted upon because they display a *reflective* character or universality that is due to their belonging to an individual *only through* their belonging to any other person. Although the individual is a "concrete person," as citizen of civil society she is also an abstract universal—she is one of the many equal individuals.[32] Her motivations are *legitimate* motivations in their selfish character because they are selfish motivations of all other individuals. In order to act as a citizen of this sphere, the individual is required to recognize this double character of her volitions—the *selfish* motivation must be recognized as a *shared selfish* motivation. Individuals "can attain their ends only insofar as they themselves determine their knowledge, volition, and action" in connection with and from the standpoint of the others, hence "in a universal way."[33] Although the subject acts on selfish motives, she acts in the name of an intersubjective "we": as Hegel claims, she must "gain recognition in [her] own eyes and in the eyes of others."[34] Individual conscience is a mirror of social feelings.

(ii) Now I come very briefly to the second issue. I have argued that on Hegel's dialectical view of action (and of the relationship between action and agent), what makes an agent 'social' is not the fact that its given reality is that of a collective or institutional subject; for, the opposite is rather the case: it is the very structure of action (or agency itself) that first institutes the agent as 'social.' As agency is realized in social structures and practices constitutive of truth, the subject that may be seen as fit for carrying it out must be a social agent; epistemic processes are not social because enforced by already given and formed social subjects. The point is clearly relevant socially, politically, and historically. For, if it is the practice that institutes the agent and not vice versa, one can consider institutions such as universities and communities of learning, museums, and news organizations as justified in their existence or in need of revision of their structure and norms on the basis of the way in which they fulfill a certain type of epistemic activity, in this case, the epistemic activity of actualized freedom. The latter not the former constitutes the criterion of truth. Herein the Hegelian example that can be offered is that of "recognition" (*Anerkennung*). In the *Phenomenology*, it is famously a certain type of practice, the action of reciprocal *Anerkennung* in the peculiar dialectical process that it displays between two individual consciousnesses that first institutes the collective subject that is the (proto-)social

"We" of spirit. It is the dialectic that develops the necessarily reciprocal action of recognition that institutes, for the first time, both the failed unrecognized individuals (master and slave) and, in the end, the successful intersubjective and eventually collective spiritual agent—and not the other way around. There are no presupposed 'agents' before and independently of the development of the process of recognition; and such process does not presuppose a certain constitution of its agents.

But I must come to my conclusion. In sum, I have proposed a concept of epistemic agency that may avoid the ambiguity and redundancy which I see inherent in its use in much of the contemporary discussion by developing it in the framework of a Hegelian social epistemology that is dialectical and based on the notion of spirit. Such epistemology is based on the idea that action constitutes the agent, and has the merit of allowing one to overcome the oppositional dichotomies between cognitive and practical, individualistic and social that in the end are responsible for the redundancy of the concept.

NOTES

1. Which has led to the denial of the possibility of 'epistemic agency': see Engel, "Is Epistemic Agency Possible?"
2. Fuller, chapter 1, section 4
3. Fuller, chapter 1, section 4. It is difficult to understand what this 'in-between' status can be.
4. This is the general suggestion of Goldberg, chapter 2.
5. I shall come back to the need of clarifying some of Goldberg's points below.
6. For a more extensive treatment of social epistemology in relation to Hegel see Nuzzo, "The Social Dimension of Dialectical Truth."
7. Goldberg, chapter 2, section 1.
8. Transcendental is the account concerned with the "conditions of possibility" of certain objective results—knowledge contents, practices, truths—insofar as such conditions are located within the knowing subject.
9. Goldberg, chapter 2, section 1.
10. John Locke, *An Essay Concerning Human Understanding* (London, 1694), II, XXVII, 8.
11. Johann Gottlieb Fichte, *Einige Vorlesungen über die Bestimmung des Gelehrten* (Stuttgart: Frommann, 1962 ff.), I, 3, 35.
12. See my remarks below at the end of this section with regard to the lack of this issue in Goldberg.
13. Goldberg, chapter 2, section 1.
14. Goldberg, chapter 2, section 1.
15. I do not see the point of Fuller's distinction between agents and persons (Fuller, chapter 1, section 2). I also have to leave here aside the question of the "normativity" connected with agency—the question is far from clear.
16. "Spirit" for Hegel is the manifold and pluralistic yet also unitary embodiment of individual consciousness and self-consciousness ("subjective spirit"), collective and institutional realities ("objective spirit"), and cultural, artistic, religious productions ("absolute spirit").
17. See Nuzzo, "Hegel's Logic as a Logic of Action" for this thesis.
18. Georg W. F. Hegel, *Phenomenology of Spirit*, §20.

19. The former is the aim of Hegel's speculative-dialectical logic and method, while the latter reflects the proceeding of the "logic of the understanding" which Hegel's criticizes.
20. See Elgin, "Epistemic Agency," 137. I am interested here only in this inversion (not in the way in which what is an "epistemically estimable belief" might be established). For a discussion of Elgin's position see also Fuller, chapter 1, section 4 (addressing what is 'epistemic' in epistemic agency).
21. See note 16.
22. See Hegel's *Encyclopedia*. §443. For a detailed account of this process see Nuzzo, *Memory, History Justice*, chapter 3.
23. Hegel, *Phenomenology of Spirit*. §177.
24. See Hegel's *Encyclopedia*. §446 Remark.
25. Hegel, *Encyclopedia*. §443, §448; §468: "property."
26. Ibid., §447 Remark.
27. Ibid., §469.
28. Ibid., §469.
29. Ibid., §469 Remark.
30. Hegel, *Philosophy of Right*. §182.
31. Ibid., §183.
32. Ibid., §187.
33. Ibid., §187.
34. Ibid., §207.

REFERENCES

Elgin, Catherine. "Epistemic Agency," *Theory and Research in Education* 11, 2 (2013): 135–52.
Engel, Pascal. "Is Epistemic Agency Possible?" *Philosophical Issues* 23 (2013): 158–78.
Fichte, Johann Gottlieb. "Einige Vorlesungen über die Bestimmung des Gelehrten," *Gesamtausgabe der Bayerischen Akademie der Wissenschaften*, ed. R. Lauth (Stuttgart: Frommann, 1962 ff.), vol. 1, 3.
Hegel, Georg Wilhelm Friedrich. *Werke* (Frankfurt a.M.: Suhrkamp): *Phenomenology of Spirit*= *Phenomenology* (vol. 3); *Encyclopedia*=Enz. (vol. 10); *Philosophy of Right*=R (vol. 7) (all are cited according to section number).
Locke, John. *An Essay Concerning Human Understanding*. London, 1694.
Nuzzo, Angelica, "Hegel's Logic as a Logic of Action" forthcoming.
———. "The Social Dimension of Dialectical Truth: Hegel's Idea of Objective Spirit," *Social Epistemology Review and Reply Collective* 2, 8 (2013): 10–25 (http://wp.me/p1Bfg0-RG).
———. *Memory, History, Justice in Hegel*. NY: Palgrave, 2012.

TEN

Epistemic Agency as a Social Achievement

Rorty, Putnam, and Neo-German Idealism

Patrick J. Reider

In *Philosophy and Social Hope*, Rorty claims that traditional realists' accounts of knowledge fail, i.e., knowledge is an accurate mental representation of existence that is free from the influence of human interest and falsifying orientations. Rorty then implies that the next available epistemic move is some form of idealism, in that there is a discernable nature and scope, as well as limits and preconditions to knowledge. He additionally denies that this idealist project is obtainable.

Rorty's rejection of both realism and idealism underpins his pragmatist view that epistemic claims can only be assessed by their social usefulness and causal restraint. While I argue that Rorty is correct to reject the traditional realist epistemic project, I also argue that his case against idealism fails. In developing this claim, I argue that *conceptual* knowledge is irreducibly based in the linguistic practices of a community of knowers, which renders even private assertions a by-product of social practices.[1] I develop this claim from within a neo-Kantian framework that borrows from Hillary Putnam's 'conceptual relativism.'[2]

The act of clarifying the above position is a necessary first step for an accurate account of epistemic agency, as a proper account of epistemic agency entails an accurate account of knowledge. In other words, just as one cannot be an honest peddler of a product that does not exist, human beings cannot be an agent of something they cannot do. In this regard, the beginning sections of this chapter can be considered the groundwork for epistemic agency, in which I argue that epistemic agency is irreducibly a social enterprise that shares the same social-linguistic preconditions as knowledge.

1. RORTY'S OPPOSITION TO REALISM AND IDEALISM

My rationale for singling out Rorty is derived from his following assertion: as a pragmatist, he does not "believe that there is a way things really are. . . . [Hence, as a pragmatist, he wants] to replace the appearance-reality distinction by that between descriptions of the world and of ourselves which are less useful and those which are more useful."[3] This type of claim is inspirational to social epistemologists, such as Steve Fuller in that he wants the social payoff of epistemic investments, even risky ones, to warrant their undertaking (see chapter 2). Conversely, Rorty's pragmatism is at odds with influential social epistemologists like Alvin Goldman and Sanford Goldberg who generally accept many of the traditional notions of truth, knowledge, and representation. In these regards, Rorty creates a convenient inroad to these divergent social epistemological views.

Why does Rorty support the standards of useful and less useful over the true and the false, which are supposedly independent of human interests? Rorty denies traditional views of truth, because he denies both that the *ideal* outcome of inquiry is possible and that humans can achieve a God-like perspective free from the contingent vantage-points of human inquiry. Similarly, he rejects the traditional 'correspondence theory of truth,' e.g., truth occurs when one's ideas or models (i.e., representations) and/or assertions concerning existence accurately depict or mirror it.[4]

Rorty's epistemic stance is based on a post-Kantian insight that occurred after the linguistic turn: any object (e.g., a 'giraffe') cannot be distinguished apart from the *concepts* and *interests* that humans develop in regard to it. In other words, one cannot get to the naked object independent of human concepts and desires. Even though he accepts that there is a "causal independence of [entities such as] giraffes from humans," he nevertheless *rejects* the view that objects can be conceived "apart from human needs and interest."[5]

At first glance, Rorty's latter claim may seem excessive. So, let's frame it with some concrete examples in order to get a better grasp of the manner in which it may be defensible. Take the existence of a mountain. The distinction between a mountain and a hill may be one of convention that concerns an accepted height, which in turn renders one geographic feature a hill and another a mountain. These types of distinctions seem manmade. What then might make a better, more 'natural' or 'found' (as in 'discovered') description of this geographic feature? Should we instead define it as having fixed, measurable, and discrete dimensions? This stance begs the question, "according to what timeframe does something have to remain fixed to be a geographical feature?" If we keep the lifespan of humans in mind, a mountain seems solid and stable. If we use the existence cycle of a star (approximately ten to eighty billion years), "mountains" are fluid and shifting like waves on an ocean.

From the vantage-point (or *intersubjective* standing) of everyday human experience, there is a sense in which the *appearance* of a mountain's *permanence* is 'objective.' Nonetheless, it is the case that our *everyday* conception of a mountain is *anchored* in our awareness and unavoidable interest in our *human* lifespans. In short, there is not some extra-temporal quality that defines traits like 'permanence,' because the observation of time is relative to the observing subject's vantage-point.

Perhaps then we should describe a mountain as a solid object made up of a certain composition of rock and soil, which in turn possesses their own atomic compositions. Yet, how helpful is this new approach? It does not seem too helpful, if we ask, is it more accurate to describe a mountain or any other object as solid and inert? Perhaps it is best to say they are made up of atoms and that the atoms are separated by significant distances compared to their own diameters. One can just as easily emphasize a similar point that atoms are conceived as having parts as well. These 'parts' are also separated by comparatively significant distances. As such, one may ask, which subatomic parts *best* describe solid objects? All of them, some of them, or the aggregates they form? It is entirely unclear how empirical science could answer this question. It seems that any given answer will be highly presumptive in nature and informed by human interest, because the question, "what constitutes the *best* description?" is not the kind of thing empirical science can study. 'Best' here concerns the *useful* and *less useful* for humans, as it concerns their interests and projects.

Perhaps objects, like mountains, are more accurately described as a seething mass of moving particles that are in a constant state of agitation—like a rainstorm, but in a highly localized region of space-time. Or is it better to argue, as "string" theorists conclude, that the smallest particles of atoms are actually tiny bands of energy? In both cases, matter as the commoner conceives it, is an illusion to macro-sized and temporally oriented beings like humans.

More relevant still, have scientists finished creating theories and providing analogies on how to conceive the world around us? Throughout the history of science, we see constant changes and shifts in the manner in which existence is conceived, modeled, and theorized. In each significantly different historical development, there simultaneously occur different socially shared desires and/or needs that drive inquiry in new directions. In turn, new approaches to inquiry can result in different ways of viewing existence.[6] The consensus or acceptance of one view or another can be seen as a response to what a *group* is willing to accept (or address), as it concerns their *interests* or meets its needs. In short, as Rorty claims, some new assertion, mode of research, or data is shown to be more 'useful' according to the interests, orientations, and projects of a community of knowers. For example, any one of the above concepts of a "mountain" could outperform the others depending on which purposes, needs, or interests direct the inquiry. In this regard, each description, in its own way, could be more 'useful' than the others or express some significant failing in comparison to its competitors.

Here, all such views of a mountain need not be incompatible. Instead, our current discourse helps illustrate the manner in which human interest holds enormous influence over the specific types of wildly varying descriptions one may employ. In this regard, our present discussion does not outright defeat the realist views that Rorty rallies against, but it does show that descriptive acts are not as transparently objective or invariable as realists such as G. E. Moore and Bertrand Russell believed at the turn of the twentieth century (a bias that still influences many philosophers). It is this type of view that motivates Rorty's following claim:

> There is no way, as Wittgenstein has said, to come between language and its object, to divide the giraffe in itself from our ways of talking about giraffes [or any entity whatsoever]. As Hilary Putnam, the leading contemporary pragmatist has put it: 'elements' of what we call "language" or "mind" penetrate so deeply into reality that the very project of representing ourselves as being 'mappers' of something 'language-independent' is fatally compromised form the start.[7]

If one is attempting to achieve some form of 'metaphysical realism,' in that one is attempting to offer a description of 'the way the world really is' *independent* of human interests and language, I could not agree more with Rorty that the distinction between 'appearance and reality' does not offer the payload philosophers hoped. (I will further support this claim in the next section.)

As I will argue, even though aspects of Rorty's rejection of traditional forms of realism are defensible, his position is too strongly stated when he asserts the following:

> In the context of post-Kantian academic philosophy, replacing knowledge by hope means something quite specific. It means giving up the Kantian idea that there is something called 'the nature of human knowledge' or 'the scope and limits of human knowledge' or 'the human epistemic situation' for philosophers to study and describe.[8]

What motivates this claim is his belief that not only is the attempt to represent the *one* true account of physical reality a mistake, but also the attempt to accurately represent *mental* existence is a "bad question" and "the root of much wasted philosophical energy."[9] As we shall see, Putnam's conceptual relativism nicely sets the stage for why the latter portion of this claim is incorrect.

2. CONCEPTUAL RELATIVISM

In *Renewing Philosophy*, Putnam prepares the way for his 'conceptual relativism' by discussing Nelson Goodman's claim that language permits numerous competing descriptions:

> [T]here are many possible choices as to what we should take a physical system to be, if we want to identify chairs and trees with physical systems: space-time region (or the gravitational, electromagnetic, and other fields that occupy those regions), or aggregates of portions of the history of various molecules. Each of these ways of speaking can be formalized, and each of the resulting formalisms represents a perfectly admissible way of speaking; but as Goodman would say (and I would agree) none of them can claim to be "the way things are independent of experience." There is no one uniquely true description of reality.[10]

Here, we have a brief overview of the type of examples I drew in the prior section concerning the manner in which one *ought* to conceive of mountains: the assessment of one's *conception* of a mountain (or any object) is relative to the interests of those evaluating it. As a result, there are many legitimate ways one can describe existence (though of course this does not entail that all forms of description are warranted or legitimate, see section 3). This type of view indicates why 'natural kinds' are never as clear cut and self-evident as the realist would have us believe. In short, once we admit several legitimately distinct ways to describe the same phenomenon, what is 'natural' about it, in that there is a *singularly* true way something exists independent of how we perceive it, is no longer clear.

Nelson Goodman argues that there is a sense in which multiple worlds exist and that incommensurate conceptual systems designate different ways in which each competing system can be legitimately construed as a different world. Putnam notes that Donald Davidson and W. V. Quine reject this claim for similar reasons:[11]

> Goodman's view that the two versions are incompatible, points out that according to standard logic incompatible statements cannot be true, and concludes that it is unintelligible to maintain that both versions are true. Even if both versions are equally good for practical purposes, I cannot say that they are both truth [...][12]

Goodman, however, believes that while "logic does indeed tell us that incompatible statements cannot both be true *of the same world*, the equal 'rightness' of both of these incompatible versions shows that they are true *of different worlds*."[13] In short, Goodman believes that incompatible views can be simultaneously held as correct, so long as such claims do not involve the same world.[14]

Putnam argues that Goodman, Davidson, and Quine are all wrong, because they all assume that descriptions from two incompatible systems must, or ought to, be translatable as having the *same meaning* in the *same* world. For example, "points are mere limits" or "points are not limits but parts of space" are incompatible ways of conceptualizing space.[15] As Putnam correctly points out, "it makes no difference to our predictions or actions which of the two schemes we use."[16] Rather than accept either the Goodman or Davison/Quine conclusions, we ought to reject their shared starting premise (from which they draw opposing conclusions): "we should simply give up the idea that the sentences we have been discussing preserves something called their 'meaning' when we go from one such version into another such version."[17] According to Putnam, there is no translatable meaning from one incompatible conceptual system into the next, even when they are "equally" useful or functionally provide the same results, because the "ordinary notion of meaning simply crumbles" when faced with incompatible schemes of existence.[18]

For the metaphysical realist, truth concerns the *one* true description of existence (though of course the same truth can be communicated in many different ways). As a result, some descriptions are clearly more accurate than others, because they are better at *representing* existence than their competitors. This view is based on the belief that there is one mind-independent way the world truly is and that some conceptual-linguistic description of it *uniquely* captures its epistemically relevant features in a way that no *in*compatible description can manage.[19] As Putnam points out, the *metaphysical realist* is *prone* to assert that such a state of affairs exists for every truth claim.

Putnam denies that only *one* true description of existence is possible, because what makes a claim or description true or false cannot be reduced to some demonstrable resemblance to the one true way in which entities exist independent of our concepts. He instead argues that numerous divergent descriptions are possible, in which their 'meanings' *cannot* be translated, without remainder, to concern the same features of epistemic interest. If this is true, then it follows that we cannot separate the conceptual/linguistic structure of descriptions, as the human vantage-point, from existence: "This does not mean that reality is hidden or noumenal; it simply means that you can't describe the world without describing it."[20] The point of Putnam's claim is not to establish some form of truism. Nor did he seek to argue against those analytic thinkers who, from 1900 to the 1960's, popularly claimed that, if descriptions and knowledge were somehow *mind-dependent* (in our current case, the use of language), then radical skepticism and immaterialism are inevitable. Likewise, he was not even interested in rebuking those who still naively do. Instead, he sought to make the metaphysical realist alive to its epistemic consequences.

In this context, *metaphysical realists* need to show that their descriptions are better than all others, and far more importantly, that they will not be challenged or replaced by better descriptions in the future. Either of these realist goals requires one to be able to show the manner in which the world is independent of the ways in which one conceptually engages it, because the *marker* of truth for realism (in general) is the way things are independent of the manner in which humans *conceive* of existence. Yet, as both Putnam and Rorty point out, *conceptual* knowledge is necessarily propositional (e.g., this is a tree, mammals cannot be reptiles, the sky is blue, etc.), without which it cannot be publicly communicated, assessed, or justified. Knowledge is thus irreducibly based on the concepts that one uses to offer assertions (in the general *form* of propositions) about existence. As a result, one's *conceptual framework* can never be separated from one's knowledge claims.

Since one's knowledge claims cannot be assessed independent of a particular conceptual framework, there are often more than one competing conceptual descriptions to describe phenomena. Similarly, since history shows us that we cannot anticipate all the ways in which a new conceptual structure will show that our current views are inadequate, we are left with the view that humans are incapable of determining which *one* description of existence is superior to all others, in all cases, i.e., one cannot know truth as it is absolute and unconditioned (i.e., without change or influence by human concepts). These insights inform Putnam's conceptual relativism: *'truth' can only be discerned when the circumstances under which one's descriptions operate and the conceptual structures in which they are formed are taken into account.*

Conceptual relativism can be characterized thus: due to the manner in which the mind functions, certain epistemic limitations can be shown — namely that approaches to truth and knowledge, which strive to determine mind-independent reality in its absolute and unconditioned state, are unattainable. If this is the case, we can, despite Rorty's claims otherwise, significantly engage philosophy that concerns 'the nature of human knowledge,' 'the scope and limits of human knowledge,' and 'the human epistemic situation.' [21] This is the case, because 'the nature of human knowledge' is propositional, the 'scope and limits of knowledge' are bound by human conceptual frameworks in which knowledge can never be absolute and unconditioned, and 'the human epistemic situation' (as we shall see) is irreducibly a social achievement brought about by linguistic norms (see section 4).

3. CONCEPTUAL RELATIVISM AND RESTRAINT

One's conceptual framework and the relations they entail are established by a conceptual schema, i.e., the structural arrangement of concepts, which establishes the accepted use and interrelation of concepts. Putnam's relativism accepts that some conceptual schemas are inferior to others in ways that Rorty's relativism does not.[22] For instance, under

Putnam's model, one can argue that the inability for a conceptual framework to make reliable or useful predictions can undermine its epistemic authority. Second, a conceptual framework that allows for new and improved discoveries is superior to one that does not aid in the development of new knowledge. Third, frameworks that are self-contradictory or fail to produce the ends one seeks in its employment are also worthy reasons to reject them. Fourth, in conceptual relativism, 'truth' is not merely what one says it is. A particular phenomenon has to be observed to fit into the criteria of the classificatory framework one holds. In doing so, it must meet the normative conditions under which it is deemed appropriate to employ specific types of descriptions.

According to Rorty's notion of useful and less useful, the above assessments may be acceptable and perhaps welcomed, *if* they appear to perform some positive social function. What makes Putnam's relativism different is that it permits us to say that they are false or highly ineffective at the epistemic level. In other words, we can make epistemic assessments about the conceptual system being used that exceeds its perceived social usefulness, because under conceptual relativism, 'truth' is still operative, though truth is irreducibly *relative* to the conceptual framework with which one forms assertions: the truth or falseness of one's assertions are only discernable within the confines of a *communicable* conceptual framework.[23]

And finally, the way things appear under the use of a particular conceptual framework can be at odds with the way the framework informs its user of how things ought to be. Take for instance, the ancient framework of the four elements of earth, wood, air, and fire. Observation shows that the theoretical principles that bind the use of these elemental descriptions are false, e.g., that wood does not contain fire (as the ancient and medieval thinkers claimed). Therefore, even if conceptual frameworks are relative to one another, effective frameworks nonetheless permit *real* (i.e., 'found') discovery, in that it is not solely a manifestation of contingent human naming and/or imagining. (Similarly, bigoted conceptual frameworks can likewise be shown to be false in that their characterization of a particular population does not hold up to observation.) In this regard, one can argue that even the 'found and made' distinction of classical philosophy can be maintained in a limited (though highly productive) way, whereas Rorty's pragmatism fails to capitalize on this insight.[24]

Once again, one may, to some degree, correctly claim that Rorty would not be opposed to the above criteria for evaluating epistemic views. He would likely argue that the principles I use for selecting one system over another hinges upon the very idea of those systems that are more useful and those that are less useful. But this account whitewashes the nuanced way in which the human mind functions, the manner in which it employs concepts, the limitations of such mental operations, and how such linguistic systems relate to observable phenomena, e.g., linguistic systems are necessarily conceptual and there are shared and often divergent normative principles governing them. The above indicates that there is in fact a way in which human knowledge operates and that it has certain limitations; namely, that one's knowledge is limited to and relative to the conceptual framework one uses as conceptual relativism claims.

According to the conceptual relativism I have been advocating, what makes something true or false is the general success or failure of one's conceptual scheme to permit subjects to observe the phenomena it names. It can effetely do so only under the conditions of its intended use, according to the norms that govern its proper use, and as it conforms or fails to conform to the phenomenon it permits the subject to notice (see section 4). When phenomena match one's conceptual framework, it is not the case that the conceptual framework just happens to be useful (as a lucky shot in the dark) or that we can just choose what is useful, as we might choose what television show amuses us the most. We can thus see considerable difference between conceptual relative structures like macro- and micro-physics, which are worthy candidates for producing knowledge claims, whereas seventeenth century suppositions concerning sub-lunar *eather* or pre-Socratic accounts of water forming the basis of all matter do not. The value of the one over the other is not merely an issue of usefulness, but more importantly their employment permit more epistemic functions.

Similarly, well defined conceptual frameworks reveal and lock certain phenomena in clear relief from one another. They may even permit empirically confirmable claims (*according to* their *own* conceptual framework and their corresponding normative principles of employment). When these types of epistemic achievements occur, one correspondingly describes a conceptual space in which the world appears a certain way to the observing subject. One can therefore carry on the post-Kantian project: when one employs a successful conceptual framework, one can make certain 'objective' claims. Such claims are *not* 'objective,' because all humans employ the same categorical framework or that there exists no competing frameworks. Instead, such claims are 'objective' insofar as individuals who correctly employ a *particular* conceptual structure (and the norms that underwrite its use) perceive and similarly describe the same phenomena. When this occurs, one can discover instances in which specific phenomena does, and if all things go well, must appear to the subject. Such success is what makes something a strong contender for knowledge and a bearer of significant truth that transcends the radical relativist's tautologies: it is only true, because that is the way one defines the object in question.

4. KNOWING AND EPISTEMIC AGENCY AS A SOCIAL ACHIEVEMENT

Rorty attempts to restrain his relativism by appealing to causality:

> Davidson's claim that a truth theory for a natural language is nothing more or less than an empirical explanation of the causal relations which hold between features of the environment and the holding true of sentences, seems to me all the guarantee we need that we are, always and everywhere, 'in touch with the world'.[25]

The problem with Rorty's causal theory of truth is that there are many different types of causes and many incompatible conceptions of causes. Likewise, the notion of cause is informed by the classifying framework in which one employs it (e.g., Aristotle's four very different conceptions of causes or the various ways people disagree as to what counts as a 'cause'). Hence, the conception of cause can be relative according to the framework in which it is employed.

Since the term 'cause' is relative to the type of conceptual framework in which it is employed and there are different types of causes floating around in different historical periods and disciplines, the term unavoidably represents different types of occurrences. However, Rorty claims that "we need to stop thinking of words as representations and to start thinking of them as nodes in the causal network which pins the organism together with its environment."[26] Yet, by what means can Rorty *avoid* the type of *representation* that "causal networks" bring to mind, in order to render his assertion intelligible? In other words, despite Rorty's belief to the contrary, causal networks, however they are defined, represent some aspect of existence, which has deep metaphysical implications (i.e., there is some implied assertion that causal networks represent existence). If we must do away with representational accounts of knowledge altogether, causal accounts are no way around this issue. It is just another form of representation that is falsely claimed not to be a form of representation.

The above sets the stage for the primary issue I want to address in this section: the main problem with Rorty's appeal to causality for epistemic restraint is that it misrepresents the social achievement of human judgment. Causal explanations of 'truth' typically lead to unsophisticated versions of externalism and reliabilism. Here, I have in mind individuals who claim that the circumstances of one's environment, independent of the individual and often independent of one's community, *cause* certain *conceptual determinations* and thereby render certain types of causal relations as a dependable means to knowledge. I will now make the case that this type of view ought to be avoided, because it undermines the active social engagements involved in judgment that enable agency and ultimately a tenable social epistemology.

Why is judgment a social achievement? Is it not I, the individual, who decides what is correct or incorrect, true or false? Do not my perceptions characterize the object in a unique way that reflects my individuality? In short, am I not solely responsible for *my* knowledge as an individual epistemic agent? The implicit assumption in these questions may be true, in the sense that human beings *ought* to be responsible for their own actions. Yet, in another sense, the above sentiments obscure the social arena in which knowledge claims are formed and function, and hence, obscure the very nature of epistemic agency.

Above, I argued that conceptual relativism is a *real* phenomenon and that this phenomenon transpires because one individual or group can use a conceptual framework in legitimate ways that counter the claims of those who use a divergent but equally legitimate framework. If this is true, we can see several important social phenomena necessary for the production and acquisition of knowledge.

First, the use of a conceptual framework is not a private endeavor. One, for instance, cannot intelligibly communicate a self-created framework of classification to others who are not competent in its use. This then implies that the *preconditions* for the *shared* assessment of one's assertions, as an individual agent, lies in the fact that the language user making an assertion is speaking to other competent language users who understand the conceptual framework being used.

Second, as Robert Brandom argues in *Making It Explicit*, discerning the correctness or falseness of one's assertions requires shared norms. More often than not, such norms, which may be expressed as accepted rules, principles, and/or expectations are unspoken and difficult to articulate, since they are typically learned through habitation rather than conscious effort. Despite the fact that we often fail to be conscious of such norms, they nonetheless underpin when it is correct or incorrect to make a claim upon which others can either condone its use or condemn its misuse.

For example, norms render us responsive to the *observational conditions* under which it is acceptable to make various kinds of *empirical observations*. To illustrate this point, consider my midnight assertion that all cows are in fact black, since that is the way they appear to me at 12 a.m. Such an assertion is wrong, because those competent in color designations abide by certain basic assumptions (i.e., norms) concerning lighting conditions. As such, even if I would say to myself that all cows really are black (even during the day), it would make little sense to me, unless I am playing some type of game with myself. Here, the appearance of cows is not what indicates the manner in which they *should* be defined, characterized, or asserted to be. Rather, the often *unspoken* normative practices with which a linguistic community *expects* and *requires* observational claims to comply predetermine how one *ought* to define or characterize observing phenomena. If one removes such normative practices from one's assertions, their use, and hence what they are intended to convey, becomes impenetrable, because such normative underpinnings function as the guiding principles by which we understand what one *intends* to communicate. Put differently, such norms function as the guiding principles of what others intend to communicate to us, in that they enable a sense of the required conditions that would render one's assertions true or false. In this regard, *all* descriptions or assertions concerning one's environment invoke acquired *normative practices* that *predispose* one to notice, frame, and define phenomena within a particular conceptual context.

In the above regard, the correctness of one's claims is always subject to preexisting norms in which even the intelligibility of private assertions hinges upon their public use. The basis of one's assertions and the manner in which they can be assessed as true or false then directly arises from one's *publicly* shared ability to *respond* and *appeal* to norms. Hence, my epistemic agency, insofar as I can be responsible for the quality of my epistemic assertions and others can hold me epistemically responsible, irresponsible, correct, or incorrect, hinges squarely on the back of the community of knowers who share the same linguistic habits as myself. It is important to note here that such norms *are not immutable or extra-communal in origin*. Instead they arise from the collaborative efforts of one's entire linguistic community, as its members defend and 'police' certain norms and seek to change and reject others.

Let us now unpack the above conclusion. The existence of more than one conceptual framework shows that the framework one uses is constructed by the community in which one learns to speak or the epistemic community one enters, e.g., the vocabulary of lawyers, astrophysicists, theologians, biologists, etc. The capacity for judgment, which underpins one's ability to make a rational choice (in that one can provide reasons for the judgments they form) and be responsible for it (i.e., the traditional notion of agent), is thus a social achievement brought about by the linguistic practices that make judgment possible. This is the case, for only through the act of learning and policing the norms that render concepts meaningful can one discern, i.e.., rationally 'judge,' if a claim or characterization is proper or improper or true or false (according to the norms in play).

Descriptions of the observable world, of mental experience, and of morality, all similarly hinge on the norms that predispose one's perception of what occurs, because they selectively render one responsive to certain conditions (and not others). In this manner, I stand with Wilfrid Sellars that all characterizations are assertions (e.g., it seems, appears, or looks to me), in that all successful applications of concepts to phenomena require judgment based upon one's access to publicly shared norms. This, however, does not mean that 'judgments' must involve deliberation, which would render the spontaneity (i.e., the seeming immediateness) of empirical perceptions neigh useless, if not impossible.[27] Judgments are instead 'active' in the sense that the *human agent* is responsible for the *act* of judgment and not some unconditioned response to external existence as uncritical forms of externalism and reliabilism claim.[28]

Here, I am borrowing from the Aristotelian claim that one is responsible for one's own character, as found in his *Nicomachean Ethics*. Aristotle argues that everyday choices are prone to become habits and that habits (when they become strong) form one's character. Character, here, represents how one is likely to act when she does not have time to consider her actions, i.e., one's unconsidered *reaction* to an occurrence. People are responsible, Aristotle argues, for their character, because they are responsible for the choices that form their habits, which in turn form their character. This is also clearly the case with epistemic judgments and habits. Do I make good choices at an everyday level as it concerns my epistemic investigations and claims, i.e., do I employ the specific norms of a given type of knowledge acquisition well, and do I appropriately reject them when they are misguided? If I do, I responsibly set myself on the road to a good epistemic character. If I do not, I irresponsibly set myself down the road of a poor epistemic character.

Even though individuals are responsible for their own epistemic views and character, they nonetheless require a linguistic community to judge, and hence, be epistemic agents. In other words, insofar as it is I who judge how something ought to be characterized, 'my' judgments (that can be publicly understood) cannot rest solely on *my* imaginings. Instead, they rest on the type of publicly shared practices, expectations, and principles, as seen in our example of color designations.

As already noted, the individual does not form concepts (of epistemic value) or their use in a social vacuum. Instead, one's social group maintains and disseminates concepts and their proper use. The individual is thus more akin to a lawyer, who merely judges what does and does not fall under the law, but rarely if ever writes such laws. And even if one does write a new law (e.g., creates new concepts, redefines the use of existing concepts and/or conceptual framework), one cannot *productively* do so unless he/she takes into account the previous laws (i.e., norms, concepts, and frameworks) and effectively argues that they fail in some sense. For example, consider Martian Luther King's redefining of equality in a racially segregated era or Albert Einstein's redefining of spacetime in the non-relative framework of nineteenth century physics. In both these examples, a paradigm shift occurred by using the norms already in place—for without a nuanced and competent understanding of a given set of norms, one cannot *knowingly* convince others to accept the *appropriate* corrective measures.[29] In this manner, norms, like laws, bind all who fall under their domain.

The above overview of the normative underpinnings of language helps indicate the manner in which the capacity to judge, and consequently act as an epistemic agent, should be framed in a neo-Kantian light: judgment is what makes the human capacity to understand possible. In short, it permits rule governed thought, and the choices concerning it, to be engaged in a reflective manner that can be argued for or against. This in turn permits the advent of rationality and conceptual knowledge.

If it is the case that we do not *passively receive* the *categories of existence*, as popularly believed throughout Western philosophy, but instead acquire concepts from long years of interaction with social groups engaged in linguistic practices, reason itself is a social achievement—an achievement that is not acquirable by those raised by apes or wolves. If the reverse is true, that one possesses concepts by being born human (Kant), through rational intuition/insight (Platonism and rationalism), or passively through empirical experience (British empiricism), the very conception of social epistemology is an unproductive farce. Why? Because if any of these aforementioned alternatives were true, there would be few (if any) social influences concerning how things *ought* to be characterized and hence understood. This would be the case, as forces that are completely independent of the social realm would predetermine how existence *should* be viewed. The historical popularity of these views indicates the reason why social epistemology has only recently been born in Western thought (outside of its cradle in German thought): philosophers have traditionally misunderstood mental phenomena like concepts, judgments, rationality, and hence the capacity to know.

5. CONCLUDING REMARKS

I have been arguing that conceptual knowledge is irreducibly a social phenomenon, because it hinges upon certain linguistic practices that can only be manifested by groups. This is not to say that one merely needs to use language to communicate what one knows; rather, I have been making the stronger Sellarsian claim that conceptual knowledge is based upon linguistic practices and renders it possible.[30]

This view indicates that norms play an even deeper role in social epistemology than Goldberg considered (in section 1, of chapter 1), for his account of norms, while insightful, does not entirely rest on the social achievement of judgment as a normative/linguistic practice. I have argued that this social achievement renders human reason possible and subsequently makes all the components of knowledge Goldberg discusses possible: namely the individual and distributed acquisition, storage, processing, transmission, and assessment of knowledge. In this regard, *all* knowledge, not just some instances of it, is necessarily one of 'epistemic dependency' on other minds. Moreover, it is this inter-dependency that grants us the status of agency, i.e., we can judge and hold others accountable for their judgments.

My account of the normative underpinnings of judgment, knowledge, and agency concerns what has been called 'cognitive integration,' in which one's cognitive status (as one who employs concepts) hinges upon a complex network of other minds, practices, and expectations, as well as a variety of social influences that affect these norms. Given more space, we could then explore how corporate bodies (e.g., governments, businesses, and religions) and technological advancements (e.g., computers, internet, and search engines) influence the normative framework that underpins the preconditions of judgment, knowledge, and agency. In brief, we could engage in a Hegelian dialectic concerning the 'Spirit' of our age, i.e., all the things that influence and generate the very conditions for human understanding, knowledge, and agency.[31]

Contrary to Rorty's claims, the phenomenon of knowledge has determinable preconditions and states that can be studied and known.[32] If this is true, there is a sense in which contemporary philosophy has rediscovered something profound about knowledge[33] that is true of even empirical knowledge obtained from the "hard" sciences: that all knowledge is equally about the human condition as much as it is about the world and that the two sides of this coin are inseparable. In other words, as Kant, Hegel, Putnam, and Rorty claim, one cannot say or know what is independent of human concepts. For this reason, I believe we are warranted to be *agnostic* about whether our human conceptions can accurately "mirror existence" and therefore forsake traditional epistemology.[34] In this vein, Rorty's pragmatism is correct, when stated in a weaker formulation: there are *many* occasions in which we *should* concern ourselves with the useful and that which productively enables optimism for the future. Fuller's echoing 'normative' approach is similarly correct: social epistemology ought to be "committed to organizing the means available to bring about or maintain some desirable state-of-affairs."[35]

NOTES

1. I believe that rationality and hence the capacity to know, in its conceptual connotation, are irreducibly linguistically acquired abilities. I also believe (as well as most of those who uphold this view) that a slew of pre-linguistic abilities play an essential role in our epistemic achievements. I nonetheless agree with Wilfrid Sellars that these abilities are not actualized in an epistemic capacity until linguistic competency is achieved (though this does not include the non-conceptual ability called 'know how' by Gilbert Ryle). Those who want to attribute basic forms of perceptual knowledge to animals and prelinguistic children can effectively do so *if* they are concerned with non-rational and non-conceptual abilities like 'know how.' I am, however, doubtful of the extent to which such achievements count as "knowledge" in the robust sense of the term. In this text, I am concerned with "knowledge" as it entails rationality. Further explications of these topics far exceed the space of this chapter and are addressed in *Wilfrid Sellars, Idealism and Realism: Understanding Psychological Nominalism*, ed. Patrick J Reider. (Bloomsbury Publishing, 2016).

2. I argue that this is the case despite the fact that Putnam believed he should have called 'conceptual relativism' 'internal realism.' Nor do I wish to suggest that Putnam is an idealist. In response to my query if he perceives his conceptual relativism and rejection of metaphysical realism as inroads to idealisms, he immediately stated, "I take my metaphysics quite serious." The implication was that as a realist, he believes one can possess significant knowledge concerning metaphysics, which presumably he does not believe idealism permits.

3. Richard Rorty, *Philosophy and Social Hope* (New York: Penguin Books, 1999), 27.

4. See Rorty, *Philosophy and the Mirror of Nature* (Princeton: Princeton University Press, 1980).

5. Ibid., xxvi.

6. Patrick J. Reider, "A 'Dialectical Moment': Desire and the Commodity of Knowledge," in *The Future of Social Epistemology: A Collective Vision*, ed. James Collier (London: Rowman & Littlefield International, 2016).

7. Rorty, *Philosophy and Social Hope*, xxvii.

8. Ibid., 34.

9. Ibid., xxiv.

10. Hilary Putnam, *Renewing Philosophy* (Cambridge: Harvard University Press, 1995), 110.

11. See Davidson's "The Very Idea of a Conceptual Scheme" and Quine's *Fact, Fiction and Forecast*.

12. Hilary Putnam, *Renewing Philosophy*, 115–56.

13. Ibid., 116.

14. What exactly Goodman means by 'more than one world' is difficult to say, but since Putnam has good reasons to reject this view, we do not need to get sidetracked by it here.

15. Hilary Putnam, *Renewing Philosophy*, 116.

16. Ibid., 116–17.

17. Ibid., 119.

18. Ibid.

19. Even 'isomorphic representations' require that some feature of existence is represented. In other words, even if one's representations are sophisticated models that hold an inexact corresponds (e.g., the universe is *like* bread pudding), rather than mirroring existence, isomorphic representations are nonetheless only considered true to the extent that they properly depict some features of existence.

20. Hilary Putnam, *Renewing Philosophy*, 12.

21. Rorty, *Philosophy and Social Hope*, 34.

22. While Rorty sometimes complains that he is called a relativist and professes not to understand what his critics mean by this label, there is a simple way it concerns his epistemic views: according to Rorty, what constitutes a good epistemic claim is *neither* its appeal to traditional conceptions of truth or accurate representations of existence. Instead, he argues that a productive social function is what makes a good epistemic claim. Here, I consider an epistemic view that 1) rejects both traditional notions of truth and representation and 2) claims that the *warrant* of a knowledge claim is primarily *relative* to the evaluation of the productive social functions it produces (as opposed to mind-independent existence) a form of 'relativism.'

23. This will ironically provide long term social benefits that would otherwise be overlooked if one only seeks overt signs of social gain. This is the case, because we cannot always anticipate what will be in the public good or how some current insight will give rise to productive social gains. If we merely look for advantages, epistemic achievements are delayed until it becomes evident in hindsight that we ought to know some particular phenomenon. In this vein, one can make the case that we should accept the rigor of epistemic investigation for its own sake and not some 'apparent' or 'bet' upon gain in the future. Having said this, I am not deaf to the irony of this latter claim. And as I will reiterate at the end of this chapter, there is much to be gleaned from Rorty's and Fuller's goal of evaluating epistemic claims in light of social gains.

24. It is not my contention that Rorty would outright reject these claims. Rather, my point is that he fails to make use of them in favor of his much stronger formulation of pragmatic relativism.

25. Rorty, *Philosophy and Social Hope*, 33.

26. Ibid., xxiii.

27. Imagine if you had to make a conscious effort before empirical experience could occur. For example, imagine if you had to make a deliberate attempt to discern the color of your grass, if there was a person in front of you or a tree, etc. Obviously, everyday perceptual experience rarely requires this *type* of active deliberation.

28. Obviously I am not claiming all forms of extremism and reliablism make this mistake. I am merely making a historical note of the sort of view that renders social epistemology an unviable discipline.

29. See Patrick J. Reider, "The Internet and Existentialism: Kierkegaardian and Hegelian Insights" *Social Epistemology and Technology: Toward Public Self-Awareness Regarding Technology*, edited by Frank Scalambrino (London: Rowman & Littlefield International, 2016).

30. This claim is made by Sellars, which he develops in "Empiricism and Philosophy of the Mind." He calls this phenomena "psychological nominalism." See Patrick J. Reider, "German Idealisms and Physiological Nominalism," in *Wilfrid Sellars, Idealism and Realism: Understanding Psychological Nominalism*. ed. Patrick J Reider (Bloomsbury Publishing, 2016).

31. Perhaps even more startling is that without this socially acquired capacity, moral agency (not just epistemic agency) is imposable as well. Without having learned concepts and the normative framework of their employment (i.e., a conceptual schema) from one's linguistic community, one could not discern, communicate, or defend one's actions as holding value over another, because he would lack the concepts to note their occurrence and correspondingly form the assertions to argue for (or against) their enactment. Of course one's thoughts may have a propositional *form* without being put into words.

32. This is the type of Idealist move I read Rorty to overtly reject in *Philosophy and Social Hope*.

33. A perspective already understood by our German idealist, Buddhist, and Hindu brethren—though perhaps in incompatible ways.

34. See Patrick J. Reider, "Psychological Nominalism, Conceptual Relativism, and Idealism," in Wilfrid Sellars, *Idealism and Realism: Understanding Psychological Nominalism*. ed. Patrick J. Reider (Bloomsbury Publishing, Forthcoming 2016).

35. See the opening page to chapter 2 of this volume.

REFERENCES

Brandom, Robert. *Making it Explicit*. USA: Harvard University Press, 1994.

Hegel, G. W. F. *Phenomenology of Spirit*. USA: Oxford University Press, 1977.

Kant, Immanuel. *Critique of Pure Reason*. New York: Cambridge University Press, 2005.

Reider, J. Patrick. "A 'Dialectical Moment': Desire and the Commodity of Knowledge." In *The Future of Social Epistemology: A Collective Vision*. edited by James Collier. London: Rowman & Littlefield International, 2016.

———. "German Idealisms and Physiological Nominalism," in *Wilfrid Sellars, Idealism and Realism: Understanding Psychological Nominalism*. Edited by Patrick J. Reider. Bloomsbury Publishing, Forthcoming 2016.

———. "The Internet and Existentialism: Kierkegaardian and Hegelian Insights" *Social Epistemology and Technology: Toward Public Self-Awareness Regarding Technological*. Edited by Frank Scalambrino, 59–68. London: Rowman & Littlefield International, 2015.

———. "Psychological Nominalism, Conceptual Relativism, and Idealism," in Wilfrid Sellars, *Idealism and Realism: Understanding Psychological Nominalism*. edited Patrick J. Reider. London: Bloomsbury Publishing, Forthcoming 2016.

Putnam, Hilary. *The Many Faces of Realism*. Illinois: Open court, 1987.

———. *Renewing Philosophy*. Cambridge: Harvard University Press, 1995.Rorty, Richard, *Philosophy and Social Hope*. New York: Penguin Books, 1999.

———. *Philosophy and the Mirror of Nature*. Princeton: Princeton University Press, 1980.

Sellars, Wilfrid. *Empiricism and Philosophy of the Mind*. USA: Harvard University Press, 1997.

Index

agency, viii, 130, 170, 174; agent-oriented, 61, 65; cognitive agency, 114, 117, 119; divine agency, 29; epistemic agency, ix, x, 10, 13, 17, 22, 25, 32, 47, 48, 54, 75, 79, 86, 92, 101, 105, 120, 148, 151, 154, 172; epistemic agents, 64; epistemic fluid agency, 131; liability model of agency, 25, 132, 140
analytic social epistemology, 44, 61
Aristotle, 28, 133, 173

Bentham, Jeremy, 22, 29, 30, 146

conceptual relativism, 165, 167
constructivist, xii, 25, 62, 148
critical social epistemology, 45

'deciding to believe', 36
Deleuze, Gilles, 131, 138
deontology, 22, 30
disagreement, 76, 80, 86
division of labor, 52, 65, 91, 93, 110

Elgin, Catherine, 32, 65, 82, 154
entitlement, 7, 22, 28, 87
epistemic: dependence, 6, 10, 128, 136; externalism, 48, 115; norms, 10, 52, 109; oppression, 128; responsibility, ix, 6, 12, 50, 51, 52, 53, 110, 115; restraint, 170
esteem, 91, 95, 102. *See also* recognition
ethical life, 150, 157
existential risk, 128
expert testimony, 75, 77, 81
expertise, 9, 11, 32, 33, 46, 62, 65, 87

Fichte, Johann Gottlieb, 148, 149, 151

Geist/spirit, 152, 154, 175

German idealism, 9, 162
Goodman, Nelson, 32, 33, 35, 165
Gramsci, Antonio, 128

Hegel, Georg Wilhelm Friedrich, 22, 23, 29, 62, 64, 148, 150, 151, 154, 175

judgment, 77, 80, 139, 170, 172, 173, 174, 175
justification, 5, 8, 44, 50, 79, 83, 114, 115, 149

Kant, Immanuel, 22, 27, 28, 29, 32, 54, 132, 146, 148, 151, 174, 175; neo-Kantian, 173; post-Kantian, 162, 164, 169

language and knowledge, 10, 16, 51, 131, 164, 165, 166, 170, 171, 173, 174; linguistic, 51, 161, 162, 166, 169, 171, 173, 174
liability, 26, 27; liability model, 27, 28, 54, 55, 63; liability rules, 24. *See also* agency: liability model of agency

Macintyre, Alasdair, 91, 100
material turn in epistemology, 47
mind, 62, 64, 111, 128, 155, 164, 167, 168; epistemic significance of other minds, 8, 10, 13, 46, 47, 48, 50, 51, 121, 146, 150, 151, 174; extended mind thesis, 111, 112; minded-ness, 43, 44, 48, 50, 56, 102, 136; mind-independent, 166, 167

normativity, xi, 15, 21, 52, 53; moral norms, 11, 13; norms, 6, 10, 17, 21, 51, 52, 56, 97, 99, 109, 110, 116, 153, 158, 169, 171, 172, 174. *See also* norms: epistemic

personhood, 22, 61, 63, 132, 136
Pettit, Philip, 96, 97, 100, 104
precautionary, 55, 67
proactionary principle, 32, 55, 56, 67
Putnam, Hilary, 10, 111, 164, 165, 167, 175

Quine, W. V., 33, 35, 165, 166

realism, 162, 164, 167, 176n2;
  metaphysical realism, 164, 176n2
recognition, 95, 99, 100, 101, 102, 105, 158, 159. *See also* Esteem
Reliabilism, 79, 86, 111, 113, 114, 116, 118, 120, 121, 170, 173

social practices, 4, 7, 15, 47, 52, 109, 147, 154, 161
subject effect, 139

superintelligence, 135

testimony, 4, 49, 52, 75, 76, 77, 86, 91
Thiel, Peter, 23, 24
transhumanism, 61
transparency, 85, 86
trust, 50, 52, 65, 78, 81, 83, 87, 136
truth, 62, 63, 77, 85, 114, 117, 120, 153, 156, 158, 162, 165, 166, 167, 168, 169, 170

utilitarianism, 22, 30, 56

virtue theory, 29, 32

warranted belief, 86
Whitley, Richard, 95, 102, 103
Williams, Bernard, 36, 92

Author Biographies

Finn Collin holds a PhD degree from UC Berkeley (1978) and a PhD degree from the University of Copenhagen (1985), where he is currently a professor of philosophy. His writings are mainly in the philosophy of science. Chief titles in English are *Theory and Understanding* (1985), *Social Reality* (1993), and *Science Studies as Naturalized Philosophy* (2011).

Fred D'Agostino is Professor of Humanities and President of the Academic Board at The University of Queensland, where he has also been Executive Dean of Arts. He co-edited the recent *Routledge Companion to Social and Political Philosophy*. A recent book is *Naturalizing Epistemology*. He is a Fellow of the Australian Academy of Humanities.

Val Dusek is Professor of Philosophy at the University of New Hampshire. He has written *Holistic Inspirations of Physics, Philosophy of Technology: An* Introduction, translated into three languages, and co-edited *Philosophy of Technology: The Technological Condition, A Reader*. He has written articles concerning the sociobiology and evolutionary psychology debates, the Science Wars, and the Hungarian Marxism of Imre Lakatos.

Sandy Goldberg is Professor and Chair of the Philosophy Department at Northwestern. He works in epistemology and the philosophy of mind and language. His books include *Assertion: On the Philosophical Significance of a Speech Act* (2015) and *Relying on Others: An Essay in Epistemolgy* (2010). He is currently finishing a book developing a social account of epistemic responsibility.

Paul Faulkner is a senior lecturer in Philosophy at the University of Sheffield. His main research interests lie at the junction of epistemology and moral psychology. He is the author of *"Knowledge on Trust"* (2011).

Steve Fuller is Auguste Comte Professor of Social Epistemology in the Department of Sociology at the University of Warwick, UK. Originally trained in history and philosophy of science, Fuller is best known for his foundational work in the field of 'social epistemology,' which is the name of a quarterly journal that he founded in 1987 as well as the first of his more than twenty books, the latest of which is *Knowledge: The Philosophi-*

cal Quest in History (2015). His most recent work has been concerned with the future of humanity.

Angelica Nuzzo is Professor of Philosophy at the Graduate Center and Brooklyn College (CUNY). Among her publications are: *History, Memory, Justice in Hegel* (2012), *Hegel and the Analytic Tradition* (ed. 2009), *Ideal Embodiment. Kant's Theory of Sensibility* (2008), *Kant and the Unity of Reason* (2005).

Orestis Palermos is a postdoctoral fellow at Edinburgh's *Eidyn* Centre. His research focuses primarily on how technology and the Web can shape our means for acquiring knowledge. Orestis also holds a BSc in Chemical Engineering from the National Technical University of Athens and he is the leader of the *Eidyn* 'Group Knowledge' pilot project.

Duncan Pritchard is Professor of Philosophy and Director of the Eidyn research centre at the University of Edinburgh. His main field of research is epistemology, and he has published widely in this area, including the monographs *Epistemic Luck* (2005), *The Nature and Value of Knowledge* (with Alan Millar and Adrian Haddock, 2010), *Epistemological Disjunctivism* (2012), and *Epistemic Angst* (2015).

Patrick Reider is a Visiting Lecturer at the University of Pittsburgh. He is the editor and a contributing author to *Wilfrid Sellars, Idealism and Realism: Understanding Psychological Nominalism* (2016) and *Social Epistemology and Epistemic Agency: Decentralizing the Epistemic Agent* (Rowman & Littlefield International 2016). He is also the Special Issue Editor for *Social Epistemology Review and Reply Collective* (http://social-epistemology.com), the on-line platform for the journal *Social Epistemology: A Journal of Knowledge, Culture and Policy*, published by Taylor & Francis.

Francis Remedios is an independent scholar. He authored *Legitimizing Scientific Knowledge: An Introduction to Steve Fuller's Social Epistemology*. He is on the editorial board of the journal *Social Epistemology* and he has published several papers on social epistemology. His current research is on social epistemology and transhumanism.

Frank Scalambrino is Senior Lecturer at the University of Akron, Ohio's Polytechnic University. He has taught graduate coursework in both philosophy and psychology. Recent publications include *Social Epistemology and Technology* (Rowman & Littlefield International, 2015) and *Full Throttle Heart* (2015). His work has appeared in *The Review of Metaphysics, Philosophical Psychology, Phenomenology and Mind, Topos, Reason Papers, Philosophy in Review, Social Epistemology Review and Reply Collective,* and the *Internet Encyclopedia of Philosophy*.

Printed in Great Britain
by Amazon